Mapping Texas and the Gulf Coast

N

MAPPING TEXAS AND THE GULF COAST

The Contributions of Saint-Denis, Oliván, and Le Maire

Jack Jackson

Robert S. Weddle

Winston De Ville

FOREWORD BY JAY HIGGINBOTHAM

TEXAS A&M UNIVERSITY PRESS : COLLEGE STATION

First edition

The paper used in this book meets the minimum requirements of the American National Standard for Permanence of Paper for Printed Library Materials, Z39.48-1984. Binding materials have been chosen for durability.

LIBRARY OF CONGRESS CATALOGING-IN-PUBLICATION DATA
Jackson, Jack, 1941–
Mapping Texas and the Gulf Coast : the contributions of St. Denis, Oliván, and Le Maire / Jack Jackson, Robert S. Weddle, and Winston De Ville ; foreword by Jay Higginbotham. – 1st ed.
p. cm.
Includes bibliographical references and index.
ISBN 0-89096-439-4
1. Cartography–Gulf Coast (U.S.)–History. 2. Cartography–Gulf Coast (Mexico)–History. 3. Cartographers–France–History. 4. Cartographers–Spain–History. 5. n-us n-mx e-fr. I. Weddle, Robert S. II. De Ville, Winston. III. Title.
GA408.5.G85J33 1990
526'.09764'1–dc20 90-36494
CIP

Contents

Maps

Foreword

Jay Higginbotham

Cartography in the early eighteenth century was not an inquiry unto itself. The discoverer's compulsion to know, to expand his conception of the world, was inextricably linked to the international rivalries of his time.

In the region north and west of the Gulf of Mexico, in that vast expanse of wilderness from Florida to Veracruz, the thrust begun by French Minister Colbert (piqued by French incursions into the Caribbean, as well as by the explorations of Marquette and Jolliet) culminated in the epic journey of Cavelier de La Salle. La Salle's mission was not only to discover, but to claim, and his discovery of a new route to the sea was less shattering to the status quo than the claims he laid in the Mississippi Valley. Moreover, his subsequent failure in 1685–87 to establish a coastal post was not so important as the wedge he had already driven between Spanish settlements in Mexico and those at Apalache and Saint Augustine. Louis XIV's expansionist aims suddenly became more focused. His attention fell on the rich mines of New Spain, and Franco-Spanish rivalry entered a more urgent phase, centering on the territory between the Mississippi River and the Río Pánuco.

The difficulty was that neither France nor Spain had any real grasp of the country in question, despite the fact that much of it had been penetrated by Cabeza de Vaca in 1529–36 and by Soto and his group's survivors in 1541–43. These adventurers had traveled widely, but none had made accurate maps, and even their written descriptions were hopelessly vague. La Salle's claim, cutting sharply through traditional Spanish territories, made a more accurate geography imperative, both for Spain and for France.

In Paris, the renowned French cartographers—Jaillot, Fer, and Delisle—were among Europe's most skilled professionals. Over the next several decades these mapmakers would publish the most illuminating *cartes du pays* yet made of the New World, casting broad light upon the dark recesses of the unknown interior. Yet, for all their skill, the French *cartographes* were in Paris, isolated in their drawing rooms, not in the wildernesses of Louisiana or New Spain. For the facts, for raw knowledge of the terrain, these draftsmen relied on the men at the scene, the explorers, none of whom were cartographers and most of whom were illiterates. Two of those who were literate and who exhibited varying degrees of draftsmanship were the legendary Juchereau de Saint-Denis (whose career in Louisiana, 1699–1744, approximated Bienville's) and the Parisian priest (reputed to have been the model for the *prêtre* of Isle Dauphine in Abbé Prevost's *Manon Lescaut*) François Le Maire.

Both Le Maire and Saint-Denis are well-known figures in Louisiana history, but their precise roles in the evolution of the region's cartography have never been fully described. Saint-Denis, a native Canadian with a lust for adventure, blazed new trails from Mobile to Texas and the Mexican capital over several decades. He showea a sharp eye for geographic detail, and even exhibited a talent for sketching. Exactly how facile his talent was no one knows, as many of his drawings are no doubt lost; but his field notes and observations of the terrain formed a wealth of direct evidence used to produce both Spanish and French maps. On more than one occasion Saint-Denis spent time in prison in Mexico collaborating with *oidor* (judge) Oliván Rebolledo, an important member of the Royal Audiencia, who was exceptionally active in developing the topography of the Río Grande country.

A more learned and skilled *géographe* was Saint-Denis's colleague at Mobile, François Le Maire. Differences in factional politics aside, Le Maire and Saint-Denis shared a practical interest in regional geography. Le Maire, though much less traveled than Saint-Denis, had important connections with the French map publishers. Aided by Saint-Denis's personal knowledge, he was able to construct maps and sketches of his own, some highly accomplished. Le Maire's larger role, however, was as a vital liaison between the rugged explorers of Louisiana and the mapmakers in Paris, the result of which made Delisle's great maps of 1718 and 1722 possible.

It is Le Maire's role, as well as Saint-Denis's, that Jack Jackson, Robert Weddle, and Winston De Ville delineate with such precision, adding to the study of mapmaking a dramatic and human element. Bringing to light the full story behind the maps was not an effortless task. Owing to the difficulty of sources (widely scattered and in various languages), research in French and Spanish colonial history has never been easy, but it is perhaps less difficult today than it was a few decades ago. In the 1950s Marcel Giraud pioneered archival research in this field, and more recently, Patricia K. Galloway and Elizabeth Shown Mills have demonstrated further the possibilities of rigorous, scholarly research.

In the same tradition, Jackson, Weddle, and De Ville have now provided a much needed and highly interesting addition to Gulf Coast history. Judged by the content, notes, and bibliography, their methodology is correct, their research exhaustive, and their contribution to Louisiana studies substantial.

JAY HIGGINBOTHAM
Mobile Municipal Archives

Mapping Texas and the Gulf Coast

The Oliván Rebolledo–Saint-Denis Maps of Texas, Louisiana, and New Spain, 1715–17

Jack Jackson and Robert S. Weddle

For a century and a half after the first Spaniard glimpsed the northern shores of the Gulf of Mexico, the territory between Apalache and Tampico lay in limbo, largely unexplored and certainly unmapped. The French intrusion into this territory, which Spain had long regarded as its exclusive domain, brought forth the first serious efforts to map it. Though the Gulf Coast had been sketched in 1519 by Alonso Alvarez de Pineda, all attempts to penetrate the interior in the sixteenth century ended in disaster.[1] Only with La Salle's descent of the Mississippi from Illinois and his subsequent attempt to establish a colony in its lower reaches did Spain stir itself to take notice of the lands briefly visited and then forgotten.

René Robert Cavalier, Sieur de La Salle, reaching the mouth of the Mississippi in 1682, claimed both the river and the lands it drained for Louis XIV of France. In 1685 La Salle returned by sea to plant a French colony on the Mississippi delta, missed his mark, and landed instead at Matagorda Bay, four hundred miles west. So alarmed was Spain at reports of this intrusion that, from 1685 to 1689, eleven expeditions were sent to find and eradicate the French menace.[2]

These expeditions produced a flurry of maps, as viceregal officials sought to understand exactly what was at stake on New Spain's northeastern frontier. Maritime searches, including a complete circumnavigation of the Gulf of Mexico, focused on the stretch of coast between Tampico and the Florida peninsula. This area, including its principal bays and river mouths, was charted by Juan Enríquez Barroto during the second maritime expedition in 1687, but the map itself is lost.[3] In 1690, Manuel Josef de Cárdenas, of the Francisco de Llanos voyage, mapped Matagorda Bay, confusingly referred to as San Bernardo, Espíritu Santo, Baye du Saint-Esprit, and Baye Saint Louis.[4] Here it was that the sixth overland expedition, led by Alonso de León in 1689, had found the ruins of La Salle's settlement. The noted cosmographer Carlos de Sigüenza y Góngora drew a map showing de León's route–doubtless using a sketch that de León had sent to the viceroy with his *diario y derrotero.* Sigüenza's map, which also bears Enríquez Barroto's coastal place names, was the first to offer real information on the Texas interior.[5]

When General de León returned the following year to establish missions among the Hasinai, or Tejas, he generated another route map, featuring an inset of San Bernardo Bay derived from the Cárdenas survey.[6] In 1691, Domingo Terán de los Ríos came to extend the Tejas missionary effort and, visiting the Caddos, drew a plan of their Red River villages.[7] But he established no new missions, and the two that had been founded near the Neches River were soon abandoned. Satisfied that the French threat no longer existed, Spain would not seriously concern itself with Texas for more than two decades.

Despite Spanish occupation of Pensacola Bay in 1698, Pierre Le Moyne, Sieur d'Iberville, early the next year gained a toehold for France at Biloxi Bay. In 1702, the French focus shifted to Mobile, whence explorers and traders ranged widely. One of these was Louis Juchereau de Saint-Denis, who was born at Quebec in 1676 and was an uncle of Iberville's wife. Saint-Denis had come to Louisiana on Iberville's second voyage (1700), commanding a company of Canadians sent to reinforce Fort Maurepas at Biloxi Bay. Two of the Talon brothers were included in his company, presumably because of their knowledge of the area. When La Salle's colony was destroyed in 1689, these two, as well as a sister and two other brothers, were taken in by various Indian tribes in Texas and were subsequently rescued by de León. They doubtless were with Saint-Denis later in 1700 when he and Jean Baptiste Le Moyne, Sieur de Bienville, journeyed up the Red River as far as the Natchitoches Indian villages, hoping to renew contacts with the Hasinai begun by La Salle and "to reconnoiter the Spanish position." With this expedition the French began a lasting trade relationship with the Natchitoches.[8]

During the next three years, Saint-Denis made several attempts to explore the Red River between the Natchitoches and the Cadodachos, but his efforts were frustrated in various ways.

These penetrations were duly reported at San Juan Bautista del Río Grande (present Guerrero, Coahuila) by visiting natives; yet there is not even a whisper to substantiate Saint-Denis's claim made in Mexico City in 1715 that he had visited San Juan Bautista ten years previously. It is evident nevertheless that Saint-Denis possessed reliable information of the region that Spain had deserted with the failure of its two East Texas missions in 1693.

In 1701 Saint-Denis became commandant of Fort La Boulaye on the lower Mississippi, near present Myrtle Grove, Louisiana. After that post was abandoned in 1707, he settled on the Bayou Saint Jean between the Mississippi and Lake Pontchartrain (now within the city of New Orleans) to trade with and maintain control over the various native groups.[9] After the War of the Spanish Succession, which ended in 1712, Louis XIV granted Antoine Crozat a trade monopoly in Louisiana. The founder of Detroit, Antoine de La Mothe, Sieur de Cadillac, soon arrived to govern the colony and make it profitable. Trade with New Spain was envisioned as the means to that end, although it soon became obvious that such commerce would have to be carried on outside the purview of Spanish law. First exploring possibilities at Veracruz, Cadillac was stonewalled, his ship not allowed to unload. If the French colony was to have any commercial exchange at all with New Spain, it must be by means of an overland route to the interior.[10]

The opportunity knocked even before Cadillac realized the necessity. It came in a letter written several years earlier by Fr. Francisco Hidalgo, seeking aid in renewing the Spanish missionary effort among the Hasinai that he and his fellow Franciscans had abandoned twenty years previously.

Prospects for trade with the Spanish settlements in Texas, wherever they might be, immediately brightened. Cadillac resolved to send an emissary in search of them. For the delicate undertaking he chose Saint-Denis, whose earlier explorations of the Red River—and perhaps all the way to the Río Grande—had given him some familiarity with the tribes mentioned in Hidalgo's letter. Saint-Denis was to trade for horses and mules and "try not to bring them back unladen."[11] Once contact was made with the priests in Texas, it was hoped that additional commerce would bring a steady stream of Spanish silver flowing into Crozat's coffers.

In pursuit of this goal, Saint-Denis twice crossed Texas to San Juan Bautista and each time found it necessary to proceed to Mexico City. Thus, he saw more of this uncharted section of the American West than any Frenchman before him. His observations of Texas, its geography, and its Indians were of particular interest to his Spanish "hosts" in the capital of New Spain, where there was no better information than the charts and maps resulting from the search for La Salle. From what the Spaniards were able to learn from Saint-Denis, they prepared new maps of the northern frontier. Saint-Denis aided the process by sketching the rivers between the Natchitoches villages (on the Red River) and the Río Grande. Concerning this drawing, Carlos E. Castañeda has written: "If this map is ever found, it will be of great interest . . . as one of the earliest maps of Texas drawn by a European from actual observation."[12]

While Saint-Denis's map has been elusive, others linked to his two interrogations have recently surfaced. Before considering the maps themselves, let us follow Saint-Denis's excursions of 1714 and 1717. On the first trip, he and his companions left Mobile early in September, 1713, their canoes bulging with merchandise advanced by Cadillac from Crozat's warehouse. They proceeded by canoe to the Mississippi—probably via Lake Pontchartrain and Bayou Saint Jean—a distance Saint-Denis gave as forty leagues; thence up the Mississippi forty leagues to the Red River; and up the Red eighty leagues to the Natchitoches villages. After establishing a depot for the merchandise, they left their canoes and traveled overland to the Hasinai villages, another forty leagues, arriving in January, 1714.[13]

Continuing in search of Father Hidalgo after some delay, Saint-Denis was accompanied by twenty-five Hasinai and only three Frenchmen: Pierre Talon and one of his brothers (most likely Robert), late of La Salle's colony, and Médard Jallot of Mobile.[14] By July, 1714, his entourage further reduced, Saint-Denis reached the Spanish outpost of San Juan Bautista. He estimated this last leg of the journey as 120 leagues.

Although the Frenchmen were cordially received by Capt. Diego Ramón, the viceroy ordered them sent to Mexico City for questioning. Saint-Denis and Jallot arrived there in June, 1715—almost a year after reaching Ramón's presidio. Gregorio de Salinas Varona, commanding the Presidio of Santa María de Galve at Pensacola, had forewarned the viceroy, Duque de Linares, about the expedition.[15] Viewing such unauthorized trading ventures as inimical to the defense of Texas and as a threat to peace with its Indians, Linares subjected the French visitors to interrogation. During this procedure, Saint-Denis gave the viceroy his small map of Texas. A formal declaration required of the Frenchmen was recorded and translated into Spanish by Gerardo Moro, "an Irishman in the viceroy's service." Moro (Morrow or Moore?), it seems, was a friend (if not an agent) of Crozat, to whom he secretly sent a copy of the testimony.[16]

This declaration included the Frenchmen's itinerary, totaling 320 leagues from Mobile, half by water, half by land. Additionally, Saint-Denis outlined the journey he claimed to have made to San Juan Bautista "ten years ago," all overland. Proceeding via the Choctaw and Natchez villages, he had found this way to be only 280 leagues, 40 leagues shorter than the Red River route.[17] He had not observed latitudes between the Mississippi and Mexico, he said, because he did not know how; he could give Moro only his estimate of leagues traveled from one point to another.

Moro, ordered to produce a map reflecting this information, delegated the task to an Englishman who resided in the viceregal court. This evidently was the map often mentioned in the *fiscal*'s report and the proceedings of the Junta General as having been prepared by Saint-Denis. The French explorer himself described the map as extending "from the Río de la Palizada [Mississippi] as far as New France [Canada]," indicating that it included more than just the Mobile–San Juan Bautista itineraries and his rough sketch of Texas.[18]

Moro cast Saint-Denis's activities in a most favorable light, emphasizing that he had come at the invitation of a Spanish priest to help win souls among the pagan Indians. Other officials, however, were alarmed that Frenchmen had penetrated to the Río Grande and were able to make accurate maps of the land traversed. Such knowledge, argued *fiscal* Espinosa, was more valuable than any acquired by Spaniards since their first missionaries had entered Texas; it could serve the French well in their illicit trade schemes, if not in expanding their conquests. Eastern Texas, he advised, must be reoccupied. Missionary activity should be resumed and a small military force sent to counteract the French influence.[19] The Junta General concurred.[20] A great irony is that, through the influence of the viceroy's pro-

French advisers, Saint-Denis was chosen to guide the expedition.

In this meeting of the Junta General of August 22, 1715, two influences converged. On the one hand was Saint-Denis, whose personal knowledge of New Spain's neglected northern regions was substantiated by a map. On the other was Juan Manuel de Oliván Rebolledo, whose comments during the proceedings impressed his fellow *oidores* (judges) of the Royal Audiencia. Oliván demonstrated an extensive geographical knowledge, embracing not only the outlying Spanish provinces of New Mexico and Nueva Vizcaya but also the regions occupied by the French. His Majesty would be well served, it was suggested, by commissioning him to make a study of the French realms in Canada and Louisiana and how they might affect the interests of Spain.[21]

While the new East Texas expedition was forming, Saint-Denis returned to San Juan Bautista and married Captain Ramón's granddaughter, Manuela Sánchez.[22] By July, 1716, the expedition was back among the Tejas, where four missions and a presidio were founded.[23] Saint-Denis, accompanied by his wife's uncle, the younger Diego Ramón, proceeded to Mobile, where Ramón engaged in a profitable sale of horses. Saint-Denis reported the rising prospects of trade to Governor Cadillac and obtained merchandise for a return trip to the Río Grande.

This second visit, in the spring of 1717, found Captain Ramón feeling pressure from his superiors and more inclined to caution.[24] He confiscated Saint-Denis's goods but allowed him to appeal to the new viceroy, Marqués de Valero. Reaching Mexico City in June, Saint-Denis perceived that French ascendancy in the Spanish viceregal court was a thing of the past.

Valero had received warnings from two sources: Martín de Alarcón, the new governor of Coahuila and Texas, and Salinas Varona at Pensacola, who had previously focused attention on Saint-Denis's contraband schemes. Salinas repeated his accusations, sending a copy of Saint-Denis's itinerary that he had obtained "extrajudicially" at Mobile, and urged that the French adventurer be arrested before further damage was done. The map from which this *derrotero* was copied, he lamented, had already been sent to France.[25] In truth, by the fall of 1716 the French Council of Marine had received from Louisiana several maps showing "the routes followed by explorers to the Spanish mines."[26]

The charges made by Alarcón were even more damning. From Saltillo he described for the viceroy an illicit trade operation so pervasive that the entire northern frontier abounded in French goods. Involved up to their ears were the "Ramonistas"—the coterie of Saint-Denis's Spanish in-laws—who must be removed from positions of power at San Juan Bautista if the traffic was to cease.[27]

Not surprisingly, by the end of July, 1717, Saint-Denis found himself in prison in Mexico City. Oliván Rebolledo, whose broad geographical knowledge of the northern regions has already been noted, was appointed to investigate. Oliván, as an *oidor* of the Royal Audiencia and a member of the Junta General, had confronted the matter of French expansion in the Mississippi Valley and the involvement of Saint-Denis on the Frenchman's first appearance in the Mexican capital in 1715. He had helped pave the way for the 1716 expedition, on which Saint-Denis served as guide and commissary. Oliván subsequently prepared a report for the king on the measures deemed necessary to "confine the French within the eastern banks of the Mississippi, which from north to south divides this Mexican America [from that of New France]. . . ."[28]

With this report went a map, perhaps the one constructed by Moro's English friend, showing the extent of the Mississippi from the Gulf of Mexico to Canada, as described by Saint-Denis.[29] In any case, Oliván was well aware of the services Saint-Denis had rendered on the 1716 expedition and his contribution to the Spanish reoccupation of Texas. In his examination of the suspicious visitor, completed on September 18, 1717, the Spanish official nevertheless asked some pointed questions concerning maps.[30]

Saint-Denis's second declaration, taken in the royal prison, reveals the earlier connection with the Irishman Moro and the Englishman who was trusted to draw maps of a region so vital to Spain. The declaration also confirms that Saint-Denis had presented the viceroy with a small map of Texas. Saint-Denis, however, denied that any map had been made showing the route from Mobile to San Juan Bautista, that is, a map that might compromise New Spain's security. He also revised his 1715 distance estimates. Oliván evidently found the more realistic figures useful when, three months later, he prepared a map that accompanied another *informe* on the problem of French expansion.[31] From all appearances, Saint-Denis was quite cooperative, his geographical assistance again extending well beyond his formal statement.

Pleased as he was with the information, Oliván may have fallen under the spell of the audacious Frenchman. On November 4 the *oidor* advised Viceroy Valero that reports of the smuggling ring on the northern frontier were exaggerated. He praised Saint-Denis's services to the Spanish king and recommended that he be released and his goods returned.[32] Then, on December 24, Oliván submitted to the viceroy his lengthy plan for thwarting French expansion along the Gulf Coast and in Texas. For illustration, he included a "more extensive" map than the two he had given Valero earlier and the one sent the king by the previous viceroy (Linares) with the Junta General's report of August 22, 1715.[33]

Thus, it seems that Oliván had made use of Saint-Denis's cartographic data from both his trips to New Spain. Whether Moro's Englishman assisted on either occasion remains in doubt, but the maps we believe to be associated with Oliván have a stylistic similarity. One of them bears a striking resemblance to Oliván's finished rendition, dated December 18, 1717, in the Archivo General de Indias (AGI), Seville, as will be seen.

The provenance of this set of seven maps is not known; neither the maker nor the date is definitely established. All seven items are of identical size, thirty-one by forty-two centimeters. The group is presently in the Archivo General de la Nación, Mexico (AGN), Sección de Gobernación, Mapoteca and is treated by the AGN as a set.[34] One of the maps (not included here) lists the Indian tribes of Texas according to Captain Ramón. This map also indicates the Spanish route to La Salle's fort and bears inside the drawing of the French fort the date 1684, which Alonso de León had found carved on the main gate. The AGN, from the skimpy evidence, has attributed the entire set to (Diego) Ramón and ascribed to each map a date of either 1684 or "Siglo XVII."[35]

This map, containing information provided by Captain Ramón, and one other of the group focus on the area between Monclova and San Bernardo Bay, without showing San Juan Bautista (not founded until 1700).[36] It seems likely, therefore, that these two resulted from the search for La Salle (1686–89).

Since the captain's role in the quest consisted only of occupying de León's post during the latter's absence from Monclova, it is doubtful that he drew any of the maps.

Two other maps of the set are incidental to the French incursion, one showing a group of villages and missions in central Mexico, the other depicting the region between Pensacola and Veracruz and containing sixteenth-century place names that were seldom used as late as 1684.[37]

It is the remaining three maps that are of interest here, for they reflect Saint-Denis's explorations as revealed by his two declarations – the very information used by Oliván Rebolledo to emphasize the French threat to New Spain.[38] These maps afford an early glimpse of the rivers between the Mississippi and the Río Grande del Norte, emerging rather tentatively, then developing with greater certainty in the final version. We see French posts lining the Mississippi River to Canada, the Great Lakes, "Río S. Lorenzo," and the Canadian capital of "Quevech." Along the lower Mississippi are the principal Indian villages frequented by Saint-Denis and his fellow traders. "Nueva Inglaterra" is poised menacingly on the Atlantic seaboard, ready to slip its bonds.

To illustrate the immediacy of these threats, Oliván gradually added the major settlements below the Río Grande. His final version – no doubt kept from Saint-Denis – shows Santa Fe's vulnerability on the upper reaches of the river and the rich mining region of Nueva Vizcaya to the southwest. This was the part of Mexico in which Crozat and his agents were most interested. Considered as a cartographic unit, these three maps open a new chapter in Spanish awareness of their shrinking New World domain – an awareness brought home by the purposeful wanderings of a Frenchman, Louis Juchereau de Saint-Denis.

The three maps reveal a progression, with improved detail corresponding to Saint-Denis's Mexico City declarations; he was obviously the major source. The map in figure 1 reflects information he gave in 1715, while that in figure 2 contains more accurate distances and additional facts on Canada from his 1717 declaration. Oliván, with New Spain's defenses in mind, incorporated into the third map the significant features of figures 1 and 2. This third map is a rough draft of the one that accompanied his *informe* of December 24, 1717, though it embraces more territory than the finished product (fig. 4) and provides greater detail in such matters as place names and provincial boundaries.[39] The relationship between the two is clear; stylistically, as in substance, they are almost identical.

The connection of figures 1 and 2 to figure 4 is less obvious. Yet the three are definitely linked by a rubric—doubtless Oliván Rebolledo's—that appears at the bottom of each, following a note to "Excelentisimo Señor." Oliván, having gathered data from a number of sources, from Sigüenza to Saint-Denis, claimed the maps as his own, without credit to either Saint-Denis or Moro and his mysterious Englishman. Indeed, Saint-Denis's name appears in Oliván's report only in connection with rumored French attempts to reach the legendary Gran Quivira.[40]

Trade routes and distances noted on figure 1 (the crudest map of the three) indicate that it was based on Saint-Denis's 1715 testimony, to which these details correspond precisely.[41] Besides the Frenchman's water-land route of 1714, another route extends upward from Mobile to the Choctaw villages thence to the Natchez on the Mississippi, and west to the Natchitoches: the route Saint-Denis claimed in 1715 to have used to reach San Juan Bautista ten years earlier. Again, the distances correspond to those given in the declaration.

Saint-Denis, by his own admission, had not visited the ruins of La Salle's fort on San Bernardo Bay and was not familiar with the coast.[42] His uncertainty as to how the rivers drained into the Gulf is reflected in figures 1 and 2, as well as figure 3, in which Oliván resorted to the maps that came from the Spaniards' quest for La Salle.

In figure 1 all the Texas rivers between the Red and the Río Grande enter the Gulf in two main trunks, both near San Bernardo, or Espíritu Santo, Bay. Such an interpretation follows the Spanish preoccupation with this bay brought on by La Salle's intrusion. Overall, the map is quite distorted, emphasizing the area of Saint-Denis's activities at the expense of its true continental perspective.

Figure 2 attempts to rectify this exaggeration, placing the relevant features on a more acceptable base.[43] From the due-south course of the Mississippi and the configuration of the Atlantic coastline, it appears that one of the stylized maps of the late seventeenth century (such as Hennepin's or Rouillard's) served as a foundation.[44] The map in figure 2 is transitional, holding to the old notion of the Mississippi flowing straight into the Gulf, yet giving the river an eastern distributary.

The northward sweep of the Texas rivers in figure 2 remains greatly overemphasized, their sources practically in the same latitude as New France. San Juan Bautista, therefore, sits almost where Santa Fe should be. In Texas, from the Red River to the Río Grande, the various distances total 259 leagues – closer to the 260 given in Saint-Denis's 1717 declaration than to the 160 given in 1715 and reflected in figure 1. The rivers are more detailed and more accurately placed than on the first map, although the lower reaches are still left to the imagination.

Texas now has the Cadodacho (Red), Sabinas, Atoyac, Asinai or Trinidad, Roxo or Espíritu Santo, San Marcos, Guadalupe, San Antonio, Medina, Hondo, Frio, Nueces, and Río Grande. The distances, taken from the 1717 declaration, are generally greater than the 1715 estimates. A note appearing at Quebec – Saint-Denis's hometown – corresponds to information given by him in 1717.[45] Completing the scope of figure 2, Oliván has added more settlements to northern New Spain. These form a virtual cordon, quite unlike their actual scattered locations over a great stretch of the frontier.

As stated, figure 3 must be considered with Oliván's *informe,* or *consulta,* of December 24, 1717, and its map, dated December 18 (fig. 4).[46] In his report, Oliván notes that, despite the measures taken since 1715, the French have extended their conquests along the Mississippi "as will be apparent from the two maps that I made for this purpose and put into Your Excellency's hands." Oliván viewed New France (Canada) and the French settlements in Louisiana as "separate countries, one contiguous to ours and the other more distant, as Your Excellency will comprehend from the more extensive map that accompanies this report. . . . " In conclusion, he refers to the "maps which I placed and now place in your hands," indicating that at least three maps passed into Viceroy Valero's possession as a result of the *oidor*'s study.[47] Thus, we conclude with some certainty that these are the maps in the AGN, reproduced here as figures 1, 2, and 3, the third being the source for Oliván's map of December 18, 1717, which made its way into the AGI.

As the key to limiting further French expansion from its Louisiana base, Oliván proposed six new presidios: among the

Tejas; on the San Antonio River; on the Cadodacho (Red) River; near the mouth of the Río Grande; at Espíritu Santo Bay, near the former site of La Salle's fort; and near the "western entrance of the Mississippi" (the maps suggest the Atchafalaya), on its west bank.[48] Symbols on figures 3 and 4 mark the locations.

To emphasize the role of the proposed Río Grande fort in protecting the Nueva Vizcaya mines and New Mexico, Oliván elaborates more on these regions than in figures 1 and 2. The mouth of the Mississippi now appears farther east, giving Texas something of its proper dimensions. Concerned that the French, by ascending western tributaries like the "Misuri," might easily reach Gran Quivira, he places this land of fabled riches at the foot of lofty mountains, a feature long popular with speculative mapmakers. He hopes that the proposed fort on the Cadodacho River will block such penetration, as well as French excursions to "Cuartelejo."[49] Suggesting that all these military operations be placed under a single commanding general, Oliván anticipates by half a century José de Gálvez's plan for making the Provincias Internas a separate jurisdiction.[50]

The excellence of the *oidor*'s study was recognized in 1719, when he was designated governor of "Pais de los Texas."[51] Although Oliván never assumed such duties, he—as *auditor de guerra* until his death in 1738—remained keenly interested not only in the geography of Texas but also in that of the entire northern frontier. As an expert in these areas, he was concerned with investigating the disastrous Villasur expedition from New Mexico in 1719 and with French designs on that province.[52] He followed closely the frontier inspection of Pedro de Rivera, 1724–28, offering his geographical insight in the planning stage.[53] Rivera's inspection produced a number of maps, by Francisco Alvarez y Barreiro, and the one of Texas appears to have been strongly influenced by the prototype seen in figures 3 and 4.[54]

While Saint-Denis may have inspired Oliván's enduring fascination with Texas, it seems clear that Oliván possessed a penetrating, inquiring mind that found exercise in fathoming the secrets of a largely unknown land and in rendering for his superiors reasonable and practical interpretations. Time and again he demonstrated qualities that filled a vital need of the Spanish colonies during his period of service.

Yet the Oliván–Saint-Denis maps were never published, for it was Spain's policy to guard closely the type of information they contained. French cartographers worked in a more open atmosphere than their Spanish counterparts. Saint-Denis's 1714 route to San Juan Bautista, for example, appears on the *Carte nouvelle de la Louisiane* by François Le Maire, drawn at Mobile in 1716.[55] Le Maire's effort coincides with the map that the commandant at Pensacola, Salinas Varona, reported had been sent to France.[56] The Marine Ministry in Paris acknowledged receipt of maps by Le Maire shortly thereafter.[57] His delineation of the Texas coast is superior to Oliván's, though it follows the same pattern. Rather than Delisle's 1703 *Carte du Mexique,* Le Maire used Sigüenza's map which was based on de León's 1689 expedition and Enríquez Barroto's diary of the 1687 voyage. The site of La Salle's Texas landing is shown, as is de León's route in 1689, and the map makes liberal use of place names from both de León and Enríquez Barroto.

Not long after the Le Maire–Saint-Denis maps reached Europe, Guillaume Delisle's *Carte de la Louisiane et du cours du Mississipi* was published (Paris, 1718), forming a cartographic image of Texas that persisted for decades. This map, showing both of Saint-Denis's trips, instantly outdated earlier Delisle maps, as well as those of his competitors. He, too, drew upon the place names that originated with Enríquez Barroto and de León, including some that do not appear on the Sigüenza map. Playing to his audience, Delisle extended the French territorial claim to the Río Grande and the Pecos River. Spain was outraged. Only a reluctance to divulge information about its North American kingdom prevented it from engaging in a cartographic war with France, such as the one that Delisle's claims in the east soon provoked with England.

Forgotten was the fact that Saint-Denis, writing to Governor Cadillac from Mexico City in 1715, had urged a Río Grande boundary for Louisiana.[58] But then it was also Saint-Denis who had schemed to bring the Spaniards back to that very region. Having excited their fears with his own daring venture, he enticed them with alluring descriptions of the Hasinai Indians and fanciful tales of their enduring love for Spaniards.[59] He then proceeded in similar fashion to manipulate his own government, by giving notice that the Spanish viceroy planned to settle La Salle's old bay and occupy Natchitoches.[60] What better way to bring Frenchman and Spaniard in close contact, that he might build an empire on contraband trade?

Indeed, Saint-Denis served his own ends better than he served either his king or his immediate lords, Crozat and Cadillac. As with so many other frontiersmen of the era, his contributions to the exploration and mapping of the Trans-Mississippi were incidental to his personal ambitions. Yet he altered the course and influenced the destiny of a large area of the present-day United States. In his collaborations with Oliván Rebolledo—the men's divergent aims notwithstanding—two exceptional minds tacitly joined in a symbiotic relationship. Their special talents united to raise the shroud that lay upon the region called Texas. Almost immediately, the results of Saint-Denis's explorations found their way onto both Spanish and French maps. Regardless of whether the maps were published, Texas was changed forever.

On the face of it, Saint-Denis seems to have achieved his objectives with remarkable success. He lived until 1744, an irritant to Spanish authorities all the while. Taking command of the Natchitoches post of Saint Jean Baptiste in 1720,[61] he was held in great respect by the surrounding Indian tribes, his Creole neighbors, his superiors at New Orleans, and even his Spanish adversaries across the Arroyo Hondo. He began, on the western fringes of Louisiana, a dynasty that lasted many years. Still, there appears a hint that his lot, and especially that of his Spanish wife, may not have been a happy one. In the closing months of his life, he petitioned the French foreign ministry to relieve him of his post, that he might retire with Manuela and their children to New Spain and live among her people. The request was denied.[62]

Louis Juchereau de Saint-Denis died at Natchitoches and was buried there, on the French side of a border he was responsible in great measure for creating. By no means does he come across as a victim of circumstances he could not control; yet his dying request suggests he would have preferred the Spanish side—or perhaps that the border had never existed at all.

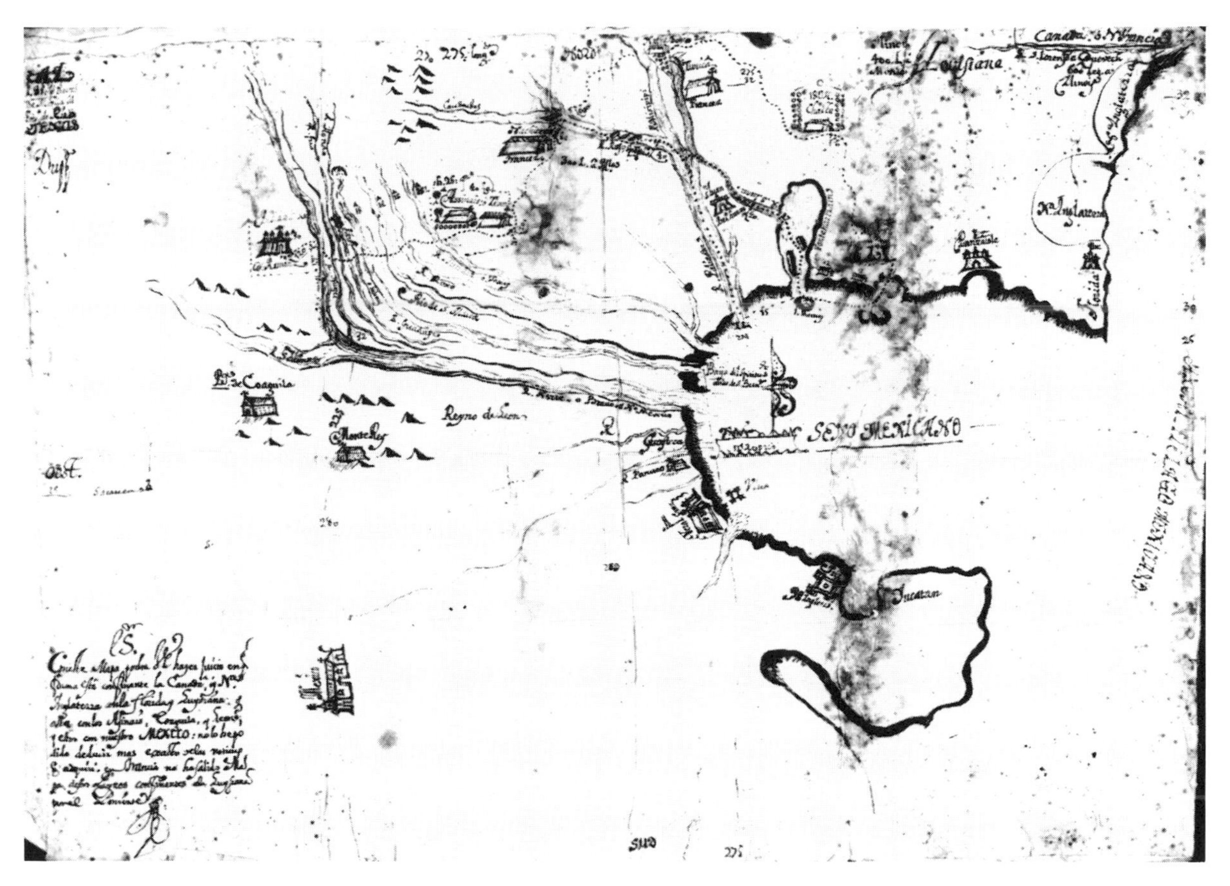

Fig. 1. Oliván's map based on Saint-Denis's 1715 declaration, made to show the viceroy the nearness of the French threat. Courtesy AGN.

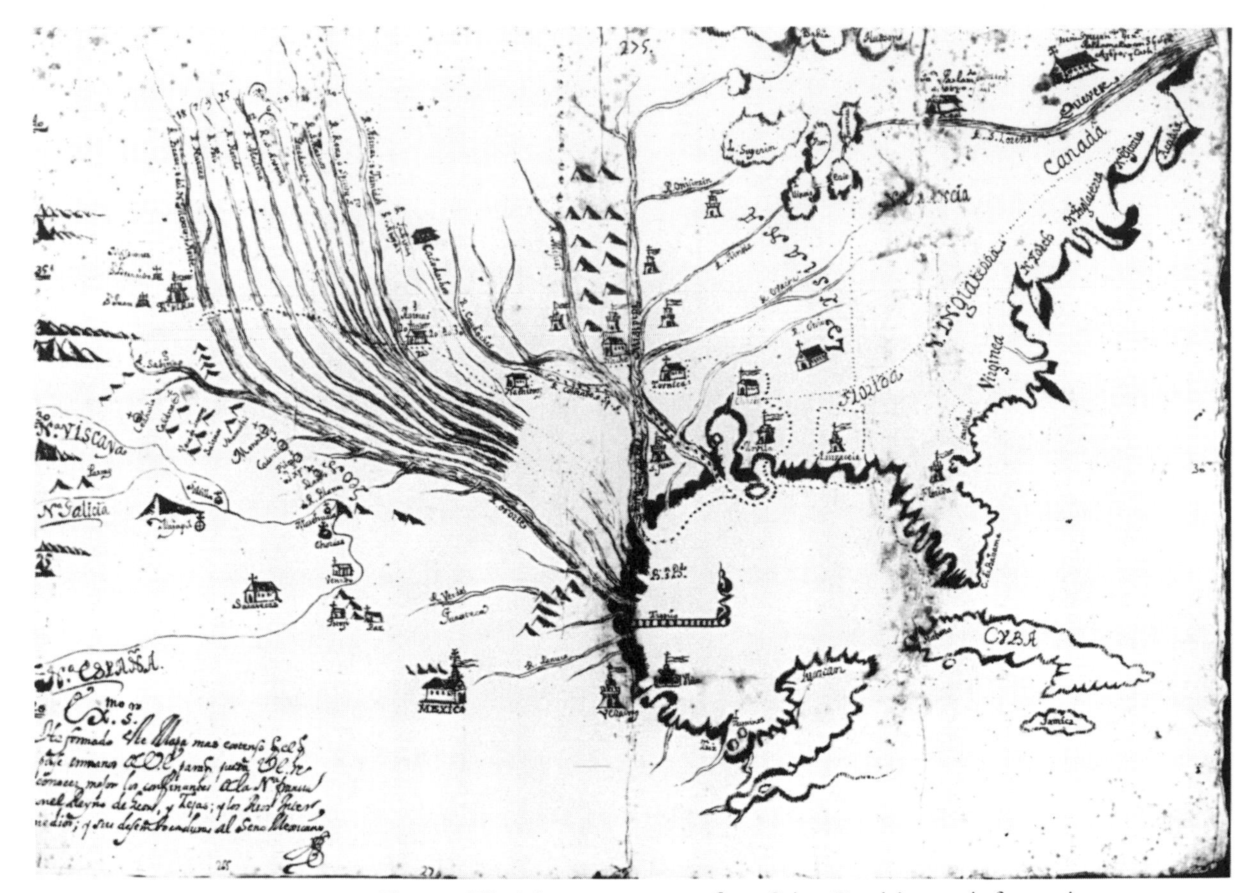

Fig. 2. Oliván's attempt to reflect Saint-Denis's 1717 information on a "more detailed" map base. Courtesy AGN.

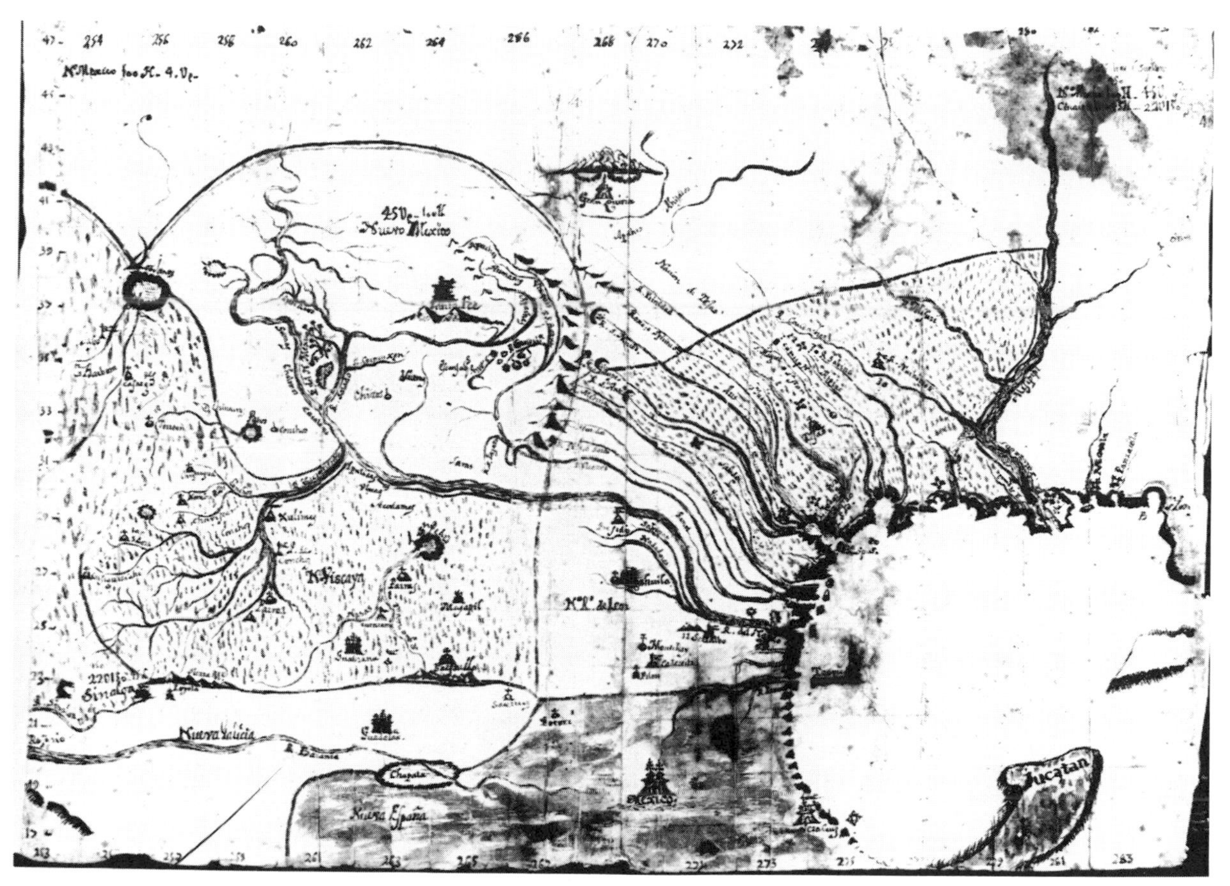

Fig. 3. Oliván's working copy for the map that accompanied his *informe* of December 24, 1717, focusing on Texas and New Spain. Courtesy AGN.

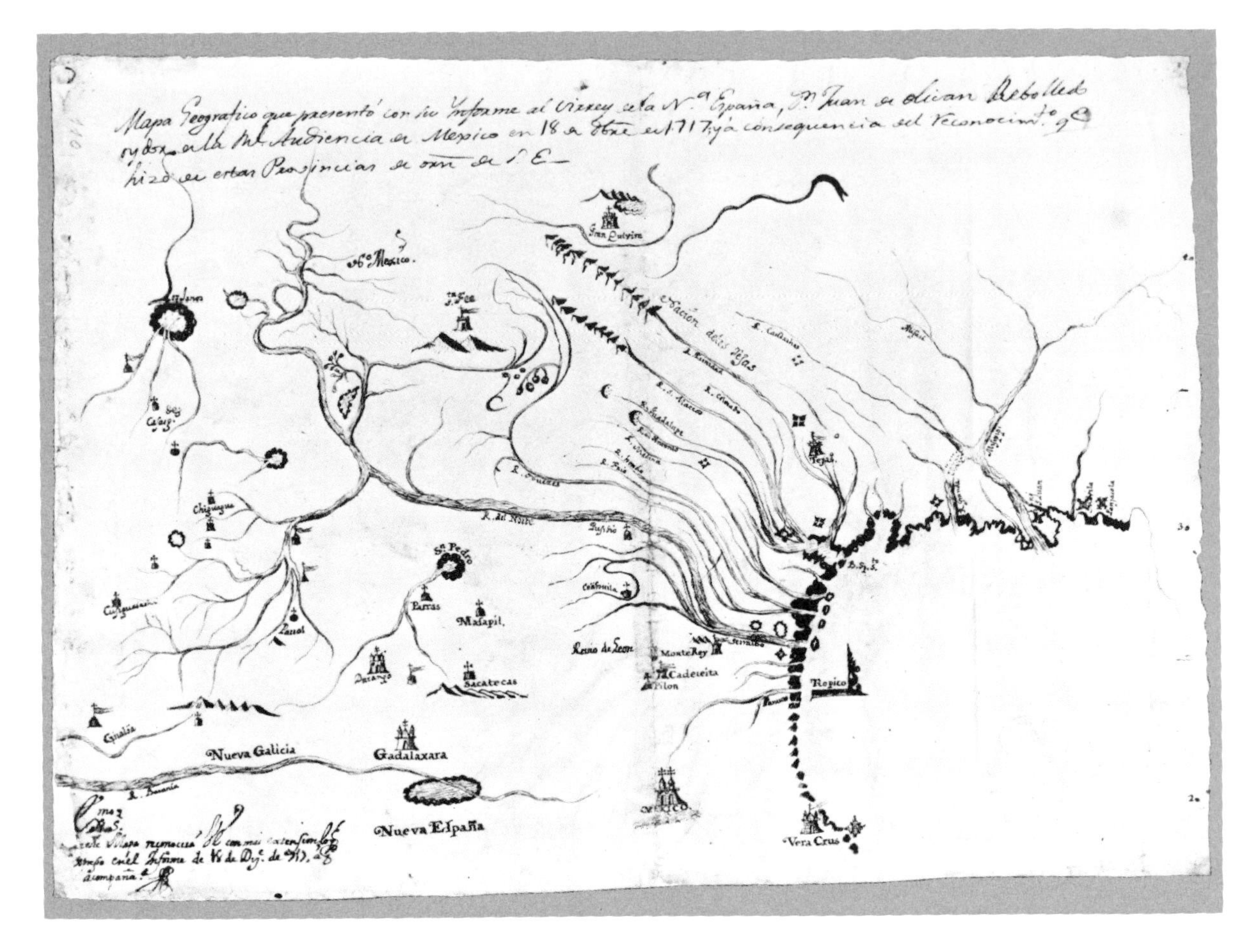

Fig. 4. Oliván's finished "Mapa Geográfico," showing his proposal for forts in Texas to protect New Spain from the French. Original in AGI. Courtesy Texas State Archives.

Le Maire and the "Mother Map" of Delisle

Jack Jackson and Winston De Ville

Just as La Salle's landing on the Texas coast in 1685 had stirred interest in this long-neglected region and had prompted a burst of mapping activity by both France and Spain, the next phase of southwestern exploration and mapping was initiated by the treks across Texas of Louis Juchereau de Saint-Denis in 1713–14, and again in 1717. Saint-Denis himself drew a map of his passage from the Red River to the Río Grande, a map that he presented to the viceroy of New Spain, Duque de Linares. It was probably little more than a crude sketch, but the map greatly alarmed the Spanish officials who saw it. Saint-Denis's map, however primitive it may have been, demonstrated a worrisome truth: the French knew more about Spain's northern territory than did the Spanish themselves.

This recognition caused the Spaniards to interrogate Saint-Denis on two occasions at the Mexican capital. His declarations and topographical information formed the basis for a series of larger maps prepared in connection with a study of the "French problem" being made by Juan Manuel de Oliván Rebolledo, a member of the Royal Audiencia. The Oliván–Saint-Denis maps represented the latest knowledge available about Texas, and about the new French settlements in Louisiana that were considered a threat to Spanish interests in the Gulf region.[1]

Even so, Saint-Denis's information had a more profound impact on the cartography of his ancestral land, France, than on that of Spain. The latter power kept the Oliván–Saint-Denis maps secret, whereas the former nation spread news of the Frenchman's discoveries through the world. Why did Spain suppress these major cartographic developments? Although part of a Spanish policy of secrecy operative since Columbus's discov-

to him by Fr. Alexandre Huvé, a Seminarian who preferred to absent himself from the fort, visiting the Indian villages around Mobile.[8] Adding to Le Maire's troubles, Louisiana's governor —Jean Baptiste Le Moyne, Sieur de Bienville—exhibited a dislike for him from the beginning. When Fr. Jacques Gravier, a Jesuit priest, visited Mobile from his Illinois mission in 1708, Gravier was asked to remain and assume the duties that Bienville accused the Seminary priests of neglecting. "There are two missionaries here from the Foreign Missions who do not do missionary work at all. . . . They give me as a poor evasion that they are afraid of being killed by the Indians. I declare to you, my lord, that these gentlemen . . . are hardly suitable for the conversion of the Indian and that very far from running to martyrdom they are fleeing from it. . . ."[9] Father Gravier reported the situation to his superiors, saying that Huvé had not learned a single word of Indian languages in his four years' residence. He described "Le Mere" as the fort chaplain "who neither chants, nor preaches, nor visits any soldiers . . . who thinks only about eating, and for whom nothing but contempt is felt."[10] As Marcel Giraud notes, such a reception does much to explain Le Maire's later resentment toward the Jesuits.[11]

The newcomer from Paris was soon exhibiting a morose attitude and a noticeable lack of zeal. "Like a lost scholar cursing the wilderness," he devoted himself to his books and the dinner table, only reluctantly officiating at religious functions.[12] By 1710, Le Maire was highly discouraged and desperate to return to France. He was convinced of the futility of the mission among the Apalaches, unhappy with his income, and bored with his unattractive surroundings.[13] Obviously he was a man of culture, and Louisiana's rough life-style proved more than he had bargained for. It is likely that his letters back home, which commenced about this time, served as an escape from otherwise dreary circumstances.

When flooding caused the main settlement to be moved closer to the mouth of the Mobile River in mid-1711, Le Maire remained at the old fort with the garrison. He then became curate of the Dauphine Island settlers before going to Spanish Pensacola in the fall of 1712, where he served as acting pastor. "The charge is vacant," wrote Le Maire, "owing to the murder of two religious whose death was the deserved punishment of the scandalous life they led."[14]

The priest had hoped his stay at Pensacola would last less than two months, but he spent almost three years there. During this time, he formed opinions of Spaniards no less caustic than those he entertained about Indians. He termed the fort at Pensacola a "land galley," garrisoned by 250 soldiers who were renowned among the Indians for their lack of courage. The civil population consisted of "scum" who had been given a reprieve from the stake, the wheel, or the rope by the tribunals of New Spain. "As you see," wrote Le Maire sarcastically to a friend in France, "I have fine parishioners."[15]

Le Maire was back in the Mobile Bay French settlements by August, 1715. His return was perhaps necessitated by the departure of Dominique-Marie Varlet, parish priest at Mobile, who left to take over the Tamaroas mission with powers of grand vicar. In his stead, Le Maire then became vicar-general of the Bishop of Quebec, to whose diocese Louisiana belonged.[16] Thus commenced for the missionary a period of intense writing and mapping. Most of his known pieces on Louisiana date from these years, as will be seen.

This outpouring of work, however, did little to raise Le Maire in the opinion of his peers in the colony. He, curate of Dauphine Island, was accused in 1716 of writing a "satirical fable" against one Sieur Raujon, who had taken up with a newly-arrived married woman in illicit fashion.[17] The following year, Marc-Antoine Hubert, the commissary-general, complained to the Council of Marine about the sad state of the colony's religious affairs, singling out Le Maire in particular:

> The indolence of the curates who make no complaints [about prostitution traffic involving Indian women] contributes to it. Besides, these inhabitants have no confidence in M. Le Maire, who performs the functions of priest here. They regard him as a petulant man, often unjust in his enmities, very indulgent toward those of whom he is fond, and . . . too much inclined to the ill-founded complaints against those whom he does not like. That obliges him to be unjust in his counsels and to support them against equity.

Admitting that Le Maire was "regular in his morals," Hubert still thought such persons "hardly suitable in a very young colony."[18]

If Le Maire's critics could be hard on him, the priest dealt out criticism just as harshly. He considered members of the local

Superior Council to be "ignorant and scandalous people." He characterized Antoine de La Mothe Cadillac (Louisiana governor, 1713–16) as a man "without faith, without religion, without honor, and without conscience." Bienville was charged with holding the "poor inhabitants" in vassalage because of his restrictive attitudes toward trade.[19] Nor was Le Maire shy about criticizing the operation of Antoine Crozat's trade concession and voicing ideas on what the company should be (but was not) doing to develop the colony.

> [Crozat] has been deceived if he imagined he would harvest before sowing, and if he based his hopes on trade with the Spaniards. Such commerce is a chimera, especially since the peace with the English. It can only be carried on furtively and by means of a shower of gold, which takes away all the profits. . . . The inhabitants believe, with regard to commerce, that they have several reasons to be dissatisfied; urged by the English, they threaten to set fire to the warehouse of the Company and to go over to the English, who come and trade as far as the banks of the Mississippi. M. Crozat must remedy this situation.[20]

How much these strong opinions had to do with Le Maire's recall is difficult to ascertain. In 1719, during the Franco-Spanish War, the board of directors of the Company of the Indies required the Seminary to bring him back to France.[21] But the company's decision (if intended to do so) did not silence Le Maire. Increasingly bitter about the return of Jesuits to Louisiana, he complained to the Council of Marine that the Jesuit missionaries sent to minister in the Illinois country were remaining along the coast. Their goal, he avowed, was to drive other missionaries from the field.[22]

The exact date of Le Maire's return home has not been determined, but it probably occured at the beginning of 1720. His departure, says Giraud, "deprived the colony of the man who, by his passage [there] and his geography, had brought together the most extensive documentation" available on Louisiana.[23] He signed the baptismal registers in Mobile for the last time on November 2, 1719.[24] In Paris, during August, 1720, he signed the marriage certificate of his niece, Marie Madeleine Le Maire, when she wed François Philippe de Marigny de Mandeville.[25] In 1721 Le Maire formally entered the Seminary of Foreign Missions, and several years later, he left France for a new missionary field, this time Siam. His ministry there was not without its controversies, but that is a story far beyond our scope. Suffice it to say that François Le Maire terminated his apostolic career at Juthia in his seventy-third year, December 21, 1748.[26] That he had remained interested in Louisiana is suggested by a notation he added to one of his maps sometime after 1744.[27]

If this extraordinary man had some personality problems, as the foregoing indicates, they did not manifest themselves in his historical and cartographic endeavors. These were the products of a trained, scientific mind, and there is little hint that Le Maire's purpose was anything other than the advancement of knowledge. His writings reveal an awareness of the common misconceptions of his day and his efforts to correct them. Le Maire's memoirs afford us remarkable insight into the problems faced by France in Louisiana, and his proposals are of constructive nature, well-reasoned and concise. Rarely do we find him dwelling on the petty feuds alluded to earlier, or even the despair that the priest felt because of personal failures. His observations take a worldview, and a positive one for the most part. Hubert was wrong: a young colony did need such persons, and we are fortunate that François Le Maire witnessed its early years.

Le Maire was concerned with all aspects of Louisiana. Beginning about 1710, his letters and memoirs are filled with interesting and informative comments on the affairs of the colony, customs of the Indians, France's neighbors on the continent, the New World's flora and fauna, its future prospects, and particularly its geography. So important were these that, had his interest been confined to geography alone, his contribution would still be noteworthy. As Giraud implies, Le Maire's catholicity was far less religious than it was scientific.[28]

The priest was also a self-confessed poet, although examples of his poetry are all but unknown. Judging from the satirical fable that so annoyed Sieur Raujon, he possessed a worldly sense of humor. But he also "amused" himself by composing "Latin hymns for all the Mysteries of Our Lord, and for the feasts of the Blessed Virgin," following the form of Horatian odes. These he planned to publish anonymously in either Mexico or France.[29]

In addition to his mapmaking talents, which incorporated a fair comprehension of the astronomical principles involved, Le Maire possessed a knowledge of botany. "I flatter myself," he wrote, "that I am the only one here to work on a natural history of Louisiana."[30] At Pensacola, "where to botanize is a question of life and death" because of Indians hostile to the Spaniards, Le Maire attempted to gather specimens for the French royal garden.[31] He claimed that he was able to understand Spanish in less than two months and speak it in less than six.[32] He could fashion sundials and he understood economics, showing a deft grasp of the measures necessary to make Louisiana a profitable colony. That he was widely read is apparent from his memoirs. His erudition is perhaps in some measure responsible for the continuing hostility to which he was subjected while in Louisiana. Jean Delanglez notes that Le Maire was "by far the best educated man in the colony," reason enough for some to resent him.[33]

Much of Le Maire's writing was directed to another learned individual, Father Jean Bobé. Bobé was a Lazarist priest of the Congregation of the Mission and one of the chaplains of the palace at Versailles. He was greatly interested in American geography. Bobé had "connections," and it was through him that Le Maire's work came to the attention of the court.[34] One of Bobé's intimates was Antoine-Denis Raudot, son of the intendant of Canada, future director of the Company of the Indies, and a member of the inner circle of Jérôme Phélypeaux, Comte de Pontchartrain, the powerful Minister of Marine. Before the turn of the century, Pontchartrain had created a special depository, a Bureau of Maps and Plans, to coordinate information about France's overseas empire. Surrounded by elite members of the cartographic world, such as Nicolas de Fer, Alexis-Hubert Jaillot, Jean-Baptiste Louis Franquelin, and the father-son team of Claude and Guillaume Delisle, Pontchartrain was keenly aware of the value of the information sent by Le Maire, which reached him via Bobé and Raudot.[35]

It was through Bobé that Le Maire's memoirs and maps also made their way into the hands of the illustrious Delisle clan.[36] Because of the great amount of scholarship that went into their maps, the Delisles were regarded as leading figures in Parisian intellectual circles of the day. With their maps, their wide correspondence, and their papers presented to learned bodies like the Académie Royale des Sciences, of which they were members, they did much to fasten their countrymen's attention on Louisiana.[37] Recognizing Le Maire as an "informant of the first order," they soon established a collaboration with him. Time and again, Le Maire was consulted on matters touching the new colony, geographical and otherwise. Thanks to the use of his information by the Delisles, Bobé, and their cadre of influential friends, Louisiana began to "outclass the older colonies in the mind of the scholarly world."[38] This fascination, of course, quickly spread to the general population.

At least four of Le Maire's memoirs are extant, though others doubtless exist, buried in French archives or circulating without being attributed to him. The first major piece is a letter written, probably to Bobé, from Pensacola on January 15, 1714. It was translated and published in 1937 by Jean Delanglez, one of the few historians aware at that time of Le Maire's significance.[39] Bobé and the Delisles made extracts of this letter pertaining to their special interests.[40]

Le Maire divided his observations of 1714 into three sections: Crozat's venture, the nature of the country, and the prospect of the missions. The second section, to which he devoted the most attention, was further divided: (1) boundaries, ports, climate; (2) plants, animals, minerals; (3) Indians, enumerating the various tribes and giving their locations, customs, and other details about them. His first section outlined a fourteen-point plan whereby Crozat's trade concession might turn a profit; the second section offered similar ideas on mining. Subsequent memoirs follow this general format, providing more details as current events unfolded and new discoveries were made. Accompanying this memoir was a map titled *Fort Louis, province de la Louisiane, et l'Isle Dauphine avec son port et rade* . . . (fig. 5), which Le Maire drew for a better understanding of the country described.[41]

Le Maire's next piece is dated March 7, 1717, at Fort Louis, prepared for the Council of Marine.[42] A copy was forwarded to the Seminary in Paris, also for the Council's use in case the original became lost. An untranslated extract of this memoir was published in *Comptes-rendus de l'Athénée Louisianais* in 1899.[43] Like the 1714 letter, this one was in response to a set of questions posed by Father Bobé and tailored to the needs of Raudot and Delisle. It is better organized and contains information on the significance of Louisiana to France's American empire, as well as Le Maire's "reflections" on the operations of the company. Such information was desperately needed by the Council in 1717, as Crozat had relinquished his monopoly and a new Company of the West was in its formative stages.[44] The section on ports and rivers reflects more knowledge of Texas, thanks to the findings of Saint-Denis. A lengthy digression on the "Sea of the West" is included, perhaps to satisfy Bobé's preoccupation with the subject.[45] New maps went with this paper, one of them showing the course of the Missouri River.

Uncertain that his second memoir had reached the Council, Le Maire prepared another in 1718.[46] Now writing with the journal of François Derbanne in hand, Le Maire expanded his 1717 remarks concerning the western country. Derbanne had accompanied Saint-Denis's 1717 return to the Río Grande, remaining there until news of Saint-Denis's imprisonment in Mexico City caused him to flee Presidio San Juan Bautista and return to Mobile. Le Maire promised to attach to his memoir a copy of Derbanne's journal from Dauphine Island dated November 1, 1717. This attachment would be included "if I have time," suggesting that the memoir was done in haste and finished at the end of 1717 or early 1718.[47] He noted that the three French trips to the Bravo had "alarmed the Spaniards terribly." Although measures were being taken in Mexico by the viceroy to bar these routes, Le Maire surmised that "from every appearance in the world, they will not succeed."[48] Thus, news of the latest western exploration of Saint-Denis and his companions was immediately transmitted to the Council of Marine and its able staff of cartographers.

The fourth of Father Le Maire's extant memoirs was written from Dauphine Island, dated May 13, 1718. It is a variation of his 1717 effort, devoted mainly to Indian customs.[49] Besides these four documents, there are passing references to other works, such as a "short relation on the discovery of New Mexico," which Le Maire had sent in 1717.[50] He also said that, in 1715, he had written an account of Saint-Denis's journey. If so, it must have been based on the letters Saint-Denis wrote from Mexico and the copy of his declaration that made its way into French hands from several sources.[51] Le Maire did not have the opportunity to speak with Saint-Denis until the latter returned to Mobile in August of 1716.

Of these yet-unfound Le Maire relations, the most fascinating is the one mentioned in a Bobé-Delisle letter of January 15, 1710.[52] After mildly reproving Delisle for not putting on his map the dotted line that separated Louisiana from Mexico farther beyond the Cenis Indians, Bobé refers to a *Manuscrite de la Louisiane* by Le Maire.[53] Bobé notes that this manuscript described a trip up the Red River by forty men in four canoes to an Indian village where Spanish was spoken and Spanish trade objects were found. "The chief of the party, M. [St.] Denis, is a native of Canada who stayed there with eight Frenchmen to push further his discoveries of those parts. . . . I am waiting for a letter from M. Le Maire; if he answers exactly all my questions, we will know a lot more about this country." Such ventures, thought Bobé, would inevitably result in an extension of the French line westward, and he urged Delisle to recognize it on his next map. That they would be successful was almost certain, because a Spanish missionary who had worked among these Indians had written to Louisiana, saying that the French would be welcomed by the Cenis. "That was the reason for this trip, which will cause us to extend our limits in this direction."[54]

It must be admitted that the events described in this 1710 letter sound suspiciously like those which took place in 1713–14. We find confirmed, at least, the activities of Saint-Denis in the region, which he perhaps visited more often than heretofore believed. Confirmed, also, is Le Maire's access to Saint-Denis's information, and the speedy transmission of it to Delisle through Bobé.

This brings us to the most significant aspect of Le Maire's work in Louisiana: his maps. Le Maire's interest in the geography of North America formed a major part of his memoirs. These he supplemented with maps, the excellence of which brought him recognition by Bobé, Raudot, and the Delisles as the "essential source" for their knowledge of Louisiana.[55] Le Maire's range of interest extended from the English settlements along the Atlantic seaboard to the Spanish colonies in Florida and Mexico. He followed the latest French developments in Canada, particularly explorations of the Missouri River, and speculated on the legendary "Sea of the West." The discovery of the latter, it was hoped, would provide a more direct link to China and Japan. Le Maire even detailed the various routes to be employed, commenting on the feasibility of each.[56] Like Bobé, he dismissed the so-called voyage of Lahontan to western Louisiana as a "fairy tale."[57] Le Maire relied instead on accounts he knew to be trustworthy, and he was not hesitant about revising his earlier opinions when new facts surfaced.

As suggested, his most valuable insights on the western country came from the explorations of Louis Juchereau de Saint-Denis. Saint-Denis had already scouted the Red River several times prior to 1706, the year of Le Maire's arrival in the province of Louisiana.[58] In addition to the trip mentioned in the 1710 Bobé-Delisle letter, Saint-Denis (in his 1715 declaration at Mexico City) claimed to have visited the Río Grande, in about 1705. He even delineated the route taken, different from the usual approach to Natchitoches up the Red River by canoe, and thence overland across Texas. Although the circumstances surrounding Saint-Denis's arrival at San Juan Bautista in 1714 make this earlier visit rather doubtful, Oliván Rebolledo credited the journey enough to plot it on his first attempt at a map of the French settlements threatening New Spain's northern frontier.[59] One thing is certain: Saint-Denis was going where no Frenchman had ventured before, and Le Maire chronicled his exploits as they were occuring.

Prior to Saint-Denis's western treks, there was little knowledge upon which cartographers could base their depiction of the region. La Salle's ill-fated landing on the Texas coast in 1685 did, however, result in several charts by the voyage's engineer, Minet.[60] As part of the effort to determine where La Salle had encroached upon their territory, the Spaniards managed to obtain copies of Minet's work in 1687, through the efforts of the Spanish ambassador to England, Pedro de Ronquillo.[61] France, likewise, soon had copies of what Spain had learned about Texas from pursuing La Salle. Abbé Claude Bernou, whose interest in North American geography is expressed in his beautiful *Carte de l'Amérique septentrionale* (ca. 1682), secured a copy of the Sigüenza y Góngora 1689 map (fig. 6), possibly from sources in Mexico.[62] It was the best that Spain possessed before Saint-Denis opened his route, and the information was used by de Fer in 1701 and Delisle in 1703. In terms of New Mexico and the flow of the Río Grande, everyone simply copied Peñalosa's map as it had appeared on Coronelli's *Le Nouveau Mexique* of 1687.[63]

This cartographic lacuna was dramatically filled when Saint-Denis returned from Mexico in 1716. In fact, news of his discoveries and route across Texas preceded his arrival at Mobile, allowing several maps to be drawn there by January of 1716. François Le Maire executed four, at least one of them tracing the route; another such map was described by Cadillac as the work of his own son.[64] As Saint-Denis did not return to the land of the Tejas, or Cenis, with the Domingo Ramón expedition until the beginning of July, reaching Mobile sometime in August, none of the new information on these maps could have been sketched at his personal direction.[65] The data, therefore, must have come through other sources—Saint-Denis's letters;[66] his declaration to the Mexican viceregal officials sent secretly to France in September of 1715, apparently by way of Pensacola-Mobile;[67] or the Talon brothers, Pierre and either Robert or Jean. Cadillac's letters to the Council suggest the Talon brothers as the most likely possibility.

The Talons, survivors of La Salle's doomed colony, possessed considerable knowledge about Texas and Mexico, doubtless part of the reason Saint-Denis took them on his journey to the Río Grande.[68] A reconnaissance of San Bernardo Bay and the Madelaine (Guadalupe) River had been conducted before the Saint-Denis venture, however, and Pierre and Jean Talon quite possibly accompanied it, revisiting the scene of their childhood traumas. This voyage, according to Cadillac, produced several maps of the easy access to New Spain inland from the coast, mentioning six Indian tribes along the way. "It is true that I have acquired all this knowledge but the relations have not yet been made," wrote Cadillac in October, 1713.[69] Only the lack of men ("Canadians are needed.") and seagoing vessels had kept him from using this route; instead, he opted to send Saint-Denis and the Talons overland, via the Red River.

It appears that the Talon brothers returned to Mobile from San Juan Bautista early in 1715 (before Saint-Denis was ordered to Mexico City), bearing letters from their leader to Cadillac.[70] They could well have carried with them a map, such as the one Saint-Denis soon presented to Viceroy Linares. Further, they may have detoured by way of La Salle's bay to determine its distance from the Spanish settlement on the Río Grande. Cadillac, in sending his son's map to the Council of Marine in January of 1716, said that the Talons had "recently" brought information that the Madelaine River was navigable a hundred leagues inland by pirogue. Even better, wrote Cadillac, the Spanish mining centers were only about seventy leagues from this region, "a fact not before known."[71]

The map that Cadillac praised so broadly (fig. 7)[72] is a rather crude affair, little better in its representation of Texas than the first attempt made in Mexico City by Oliván Rebolledo. It does, however, show Saint-Denis's route from the village of the Yatasses to San Juan Bautista.[73] Adding to the distorted east-west orientation of the Texas rivers, several are listed out of sequence. What was evidently intended as the Nueces is labeled "St. Antoine," and "Bagres" (Spanish for a variety of freshwater fish) is given to the stream that most contemporary maps show as the Hondo or Frio. The course of these rivers below Saint-Denis's route shows that the French were as ignorant as the Spaniards of their correct drainage into the Gulf, not saying much for the Talons' grasp of geography.

In pinpointing Boca de Leones, however, "where there are rich merchants who have plenty of piastres and . . . [silver] ingots" unmarked and available for trade, Cadillac's letter answers the big question that had motivated the French expeditions to New Spain. Cadillac proudly noted that the distance to the mines by way of the Madelaine ("called by the Spaniards the Guadeloupe"), was much shorter than by the other road.[74] On the younger Cadillac's map, it certainly appears to be a more direct route, due to the compressed coastline between Mobile and San Bernardo Bay, and the exaggerated length of the Guadalupe, which flows parallel to the Río Grande and very near it. Thus, two routes were available for exploiting the market provided by these long-sought mining centers: overland, via the Red River, and by sea to La Salle's bay, thence inland up the Madelaine.[75] As the Talons—not Saint-Denis—were the "experts" on the latter route, it is likely that the governor's map for the Council was primarily based on information gleaned from them.

That map must be regarded as a mere curiosity compared with the four maps that François Le Maire sent to the Council of Marine that same month, on January 25, 1716.[76] The Talons (or information they brought from Saint-Denis) possibly contributed to these maps, but Le Maire must have relied on other sources in preparing them. These sources, as inspection reveals, were far superior to those used by the governor's son. Not received by the Council until late August—around the time Saint-Denis finally returned to Mobile—Le Maire's four maps were described as showing "the country and coasts of Louisiana, and of the newly-discovered routes for going by land to New Mexico [Texas], and for reaching the most distant western and northern [Indian] nations."[77]

The Council would have received Le Maire's maps earlier had they not been sent to Father Bobé, who evidently kept them for a while. On August 17, Bobé wrote Delisle from Versailles: "If you desire I will let you have the maps which Le Maire addressed to me. I made copies before sending them to M. Raudot [who brought them to the Council's attention]. They will give you much light as to the inland country, the rivers, the localities of the Indian nations, the coast, and the peninsula of Florida."[78] With these maps a ray of light did, indeed, begin to penetrate the region described, an illumination that was to dazzle the world with Delisle's 1718 map of Louisiana. Le Maire's contribution to this cartographic chain, however, was only beginning. Other maps were to follow these, for which the Council thanked him on October 28.[79]

Of the four maps received by the Council of Marine in 1716 – presumably Le Maire's originals sent to Bobé, which the latter copied before giving them to Raudot – only two have been identified. The most important is titled *Carte nouvelle de la Louisiane et païs circonvoisins dressée sur les lieux pour être presentée a S. M^te^. T.C. par F. Le Maire pretre parisien et mission^e^. apostolique MDCCXVI.* (fig. 8).[80] Another version of this map, prepared by a skilled draftsman – possibly within the ministry and for its use – is also extant (fig 9). It is only a more finished product, having a few minor additions but no alterations of Le Maire's material.[81]

The other known 1716 map is untitled, except for the notation "Delineabat F. Le Maire P.P. missionaire Apostolic anno 1716" (fig. 10).[82] It shows the coastline, beginning below La Salle's bay and extending to the Florida peninsula. It is doubtless one of the "two other" maps to which Le Maire refers in a notation on his larger map, saying that they should be consulted for a more exact understanding of the lower Mississippi, its delta and lakes.[83] This notation was retained on the polished copy of Le Maire's *Carte nouvelle.* On the coastal map, we find a legend used by many subsequent mapmakers: "All of this coast west of the Mississippi is inhabited by nomadic cannibals."

Until they are found, we can only speculate about the subject matter of the remaining two maps. According to the comment on Le Maire's larger map, one of them showed coastal features in greater detail, possibly focusing on Pensacola, Mobile, or the Mississippi delta. Le Maire had earlier drawn maps of the bays where the French and Spanish settlements were situated, and Delisle's map contains an inset of the delta and surrounding coast. Whether this inset relied on a missing Le Maire sketch cannot now be determined, but his information was certainly the best available at the time Delisle's map was published.

Among the collection of Delisle sketch maps at the Archives Nationales, there is a Le Maire map titled *Coste de la Louisiane* (fig. 11) that very well fits the description for one of the missing 1716 maps. It shows the delta in detail, extending eastward beyond Pensacola Bay and northward above the mouth of the Arkansas. This map, however, is dated 1718, and it contains an elaborate rendering of the Red River as far west as the settlements of Natchitoches and "Adaies," the latter noted as Spanish. As these developments occured after 1716, Le Maire's *Coste de la Louisiane* seems to be properly dated, 1718. At most, we can guess that it was a refined version of an earlier (1716) map, done after more information was obtained from Saint-Denis or Derbanne concerning the Red River.[84]

The fourth 1716 map may have dealt with the Missouri River. Judging from remarks that Le Maire made in his memoir of March 7, 1717, however, it is more likely that this map went to France at that time, not as part of the 1716 shipment. In discussing the question of a linkage between the Missouri and the upper Bravo (Río Grande), he says:

> The sources of the Missouri are still unknown but there are strong reasons to believe that they are not far from the place where the Rio Bravo leaves [the Missouri], and it very much appears that the fork of the Bravo is at the location that I have marked in the note placed beside the Missouri, which must be followed in putting this map in fair copy. . . . Follow the map of the Course of the Missouri and correct my large one [*Carte nouvelle?*].[85]

In such a manner did Le Maire instruct the leading cartographers of his nation exactly how to draw their maps. That they heeded his advice and put his rough sketches into "fair copy" is evident from a study of Guillaume Delisle's work in particular.

Delisle's use of these 1716 maps is established by two of Bobé's early 1717 letters. In the first (?) he states: "I am overjoyed that you found something in M. Le Maire's maps that will assist you in making a good map of Louisiana."[86] Writing again on March 8, Bobé says:

On 1 March I received the map that you returned. I am sending three others that I hope, Monsieur, will please you more. . . . I believe the error that you noticed relating to the degrees was not M. Le Maire's, but that of [the copyist], in a great hurry, having been pressed by M. Raudot to return the original. . . . I can assure you that the map of the Gulf of Mexico, and that of the coast of Louisiana, which was copied by the same [person?], conform very much to the originals. I verified them carefully. That of Louisiana and of the "route of etc." I copied myself with great exactitude. . . . I am impatient to see the corrections that you made on the map [of Louisiana] I sent you.[87]

Bobé is clearly referring to Le Maire's 1716 *Carte nouvelle* and the accompanying maps that he had promised to let Delisle use. What became of the copies made by Bobé and Delisle has not been determined, but it can be seen that at least four versions of the *Carte nouvelle* once existed.

Le Maire's 1716 maps, coming virtually out of the blue, must have created a sensation in France. How could he possibly have drawn maps unlike any before them, maps that would forever change the world's notion about the land depicted? The truth, of course, is that Le Maire studied the work of other mapmakers (including Delisle), using what little they had learned over the centuries, and then corrected their charts as much as he could with the new information available to him.[88] The strength of his effort lies somewhat in the weakness of his precursors, for Le Maire was largely filling in blank spaces and redrawing fictional features.

One map to which Le Maire definitely had access was Sigüenza's 1689 route map of the Alonso de León expedition (fig. 6). As noted, Abbé Bernou had secured a copy of this map, and French cartographers starting with de Fer began to incorporate elements of it into their maps. Thus did the route taken by the Spaniards to find La Salle's fort at Baye Saint Louis begin to grace French maps (fig. 13).[89] Le Maire, however, worked directly from the Sigüenza map, not the simplified versions published in France. His 1716 maps show all too well his reliance on a facsimile of the original for this section of Texas, and they contain information that does not appear on secondary sources, such as de Fer's maps.

Le Maire's Texas coastline and San Bernardo (or Saint Louis) Bay, for example, are practically identical, as are the place names thereon. Like Sigüenza, he sketches one of La Salle's vessels (the *Belle*) wrecked inside the bay. The flow of the rivers from the interior to the coast is likewise consistent, as are their names. Le Maire traces de León's route with the same precision as Sigüenza, showing the same three Indian villages visited at the end of the trek. He has the Nueces as a tributary of the Río Bravo, a misconception that would not be corrected for several decades. Below the Bravo, we also see Sigüenza's "Cuahuila" nestled in its river network, an aspect that Le Maire reproduces faithfully–except to spell it "Cahouila."[90]

How and when Le Maire obtained this map is uncertain, but he could not have drawn his *Carte nouvelle* without it. Apart from the convincing graphic evidence, we have Le Maire's own words: "I have already sent and will send again today the different points of the route the Spaniards took from Cahuila to this Bay."[91] A table containing such information adorns Sigüenza's map, and Le Maire clearly had a copy in his possession.

As for the coastline beyond La Salle's bay, and the rest of the continent, the French were rapidly gathering knowledge. Le Maire had only to revise existing maps, based on the most recent explorations from Louisiana and Canada–information that had not yet made its way into published maps. But how well Le Maire managed to assemble all these contradictory fragments into a coherent whole is the remarkable thing. To his credit, he eliminated the useless vestiges that so many other cartographers had felt obliged to perpetuate, compensating for their lack of knowledge. Thus, his maps were rendered in the lean style that was to signal a new era of scientific mapmaking. Though not involved in exploration himself, he was in the right place at the right time to secure firsthand information from the actual explorers. Further, Le Maire's learning and talent enabled him to convert the raw observations of frontiersmen into scientific terms, thereby adding significance to the accomplishments of pathfinders like Saint-Denis.

While praise for Le Maire's maps was coming from France, the ill tidings of these developments were being emphasized in Spanish circles. Le Maire, it will be remembered, had spent three years at neighboring Pensacola, during which time he gained an intimate knowledge of the Spanish settlement. There, in 1713, he drew a map of the coast between Mobile and Pensacola, containing an enlarged plan of the fort. This map (fig. 12)[92] was sent to Spain but Le Maire doubtless kept a copy. His 1714 letter mentions that it was his habit to dine with the quartermaster and the "governor," the latter being Gregorio de Salinas Varona, who commanded Presidio Santa María de Galve.

Le Maire must have found Salinas an interesting dinner companion, for the commander had broad experience on New Spain's northern frontier. In 1691, he was serving as captain of the Monclova presidio and commanded the troops aboard the sea division of Domingo Terán de los Ríos's expedition to Texas. He landed at Matagorda Bay, examined the ruins of La Salle's fort, and explored the estuary. Two years later, he led an overland expedition from Monclova, carrying supplies to the impoverished East Texas missions. Salinas remained in Coahuila as acting governor of the province until 1698.[93]

Years later, in 1717, after Salinas had been transferred to Pensacola from Honduras, he was remembered by Oliván Rebolledo as the man "most fit" to make a thorough reconnaissance of the Texas coast because he "has been in this Bay [of San Bernardo], going there both ways, over land and sea, and because he inspected the place where the French had their village and other places . . . equally or advantageously better suited for fortification."[94] Oliván wanted Salinas to command the pivotal presidio that he planned to build on La Salle's bay, thereby securing it from the French. Had Gregorio de Salinas Varona been the slightest bit talkative on the subject of geography, François Le Maire could have asked for no better dining arrangement than the one he enjoyed at Pensacola.

It is reasonable to assume that the priest maintained relationships formed at the Spanish post, even after returning to his clerical duties at Mobile Bay in 1715. Perhaps it is more than coincidental that Le Maire was able to draw a map of Texas shortly thereafter, leading to the strong suspicion that he obtained a copy of the Sigüenza map from the commandant of Pensacola. Had not Le Maire lent his mapmaking talents to the service of Spain? Was he not caring for the spiritual needs of the garrison, as well as making useful things like "huge" sundials to regulate the fort's military drills? Certainly, no Spaniard of Le Maire's acquaintance knew more about Texas than Salinas, or was more apt to possess a draft of the map in question. No one, of course, need ever know how the copy had been obtained.

Such off-the-record exchange of topographical information—even between international rivals—was nothing new, as anyone who studies the subject will readily comprehend. Saint-Denis, after all, had given the Mexican viceroy a map of his Texas exploration and aided Oliván Rebolledo in preparing maps of larger scope. Gerardo Moro, who took Saint-Denis's declaration at Mexico City, promptly sent a copy of it to France, along with a map showing "the nearness of the Canadians to New Mexico and other Spanish lands by the route marked thereon."[95] Cadillac, in reporting a similar situation to Pontchartrain in 1713, expressed very well the prevailing mentality. The two maps that revealed the route to the Spanish mines by way of San Bernardo Bay, he said, had been acquired under promise of "inviolable secrecy."[96] Cadillac knew that Minister Pontchartrain did not care about his source (obviously Spanish); that France now had the maps was all that mattered.

If little that transpired at Spanish Pensacola was unknown to Le Maire, evidently the same applied to Salinas about affairs at French Mobile. The circumstances of how he did it are not spelled out, but Salinas managed to gain access to Le Maire's map. Writing both the king and the viceroy early in 1717, Salinas sent a *derrotero* (itinerary) obtained "extrajudicially" from the map before it had been sent to the French court.[97] Like Cadillac, Salinas protected his benefactor while fulfilling his duty to his sovereign.

The itinerary of Saint-Denis's journey forwarded by Salinas mentions the following rivers: Red, Trinity, Colorado, San Marcos, Guadalupe, León, San Antonio, Profundo (Hondo), Nueces, and Bravo—all of which appear on Le Maire's 1716 *Carte nouvelle*. As only six of these rivers are cited by name in Saint-Denis's first declaration, and hardly any of them were accepted by French mapmakers prior to his trip, it is evident that Salinas still had contact with his former priest at Pensacola. However he (or his agents) obtained the information, there is little doubt that Le Maire's map was the source for it. Further, Governor Salinas warned his superiors that Saint-Denis, Derbanne, and others had left Mobile on November 15, 1716, for a return trip to Mexico, taking with them 30,000 pesos' worth of contraband goods.[98]

This news greatly complicated what Saint-Denis hoped would be a successful trading venture. In fact, it led to his arrest and the confiscation of his merchandise once he reached San Juan Bautista. Salinas's revelation that the French court now had a map of his route to New Spain caused Saint-Denis more grief when he faced his inquisitor at Mexico City, Oliván Rebolledo. Saint-Denis denied that he had made any such map,[99] but it was obvious to the Spaniards that their northern frontier had been greatly imperiled by his journeys. Only Saint-Denis's willingness to share his information—and Oliván's recognition of his services during the 1716 reoccupation of Texas—saved the daring Frenchman from being shipped in chains to Spain.[100]

While Saint-Denis was facing an uncertain future in the prisons of Mexico, Le Maire was writing more memoirs and drawing more maps in Mobile, utilizing his information. The memoir of March 7, 1717, directed to the French Council of Marine was accompanied by "several" maps.[101] Bobé, writing Delisle on October 16, 1717, refers to the maps that were expected to accompany a copy of the memoir being forwarded to the Séminaire des Missions Étrangères:

> M. Le Maire is also sending them [the Seminary] some maps, much more precise than those of last year [1716], because he based them on his memoirs that he lost and [then] found. These maps, as well as the memoir, should be presented to the Council [by the Seminary]. M. Le Maire sent me the rough draft of these maps, fearing that the Council would not communicate them to me.[102]

It is to be presumed that Delisle worked from these 1717 charts, adding to the information gained from those of 1716, as Bobé was urging him to return Le Maire's "rough drafts" in March of 1718.[103]

None of these "much more precise" 1717 maps has been located. As noted, one traced the "Course of the Missouri" and was to be used to correct Le Maire's larger map—either his *Carte nouvelle* or an improved version executed the following year. Certainly the title of the 1717 memoir suggests that the map being forwarded was a significant accomplishment and not merely a regional sketch. The only map that remains attached to the memoir is an engraving by Nicolas de Fer—basically a copy of Delisle's 1703 *Carte du Mexique*—which Le Maire apparently returned with suggestions for improvement.[104] What these suggestions encompassed is unknown.[105] The title of de Fer's 1718

Le Cours du Missisipi ou de St. Louis gives credit to Le Maire's memoirs, but it is evident that de Fer did not pay much attention to them or bother to alter the misconceptions he had earlier pilfered from Delisle's map.[106]

Fortunately, Guillaume Delisle showed no such reluctance, nor did another mapmaker known only to us as Sieur Vermale. Both readily absorbed Le Maire's cartographic innovations and supplemented them by extracting details from his writings. Let us first consider Delisle's *Carte de la Louisiane et du cours du Missisipi* (fig. 14).[107]

Perhaps no other map of the eighteenth century has received the acclaim lavished on this production, and rightly so. Kohl calls it "the mother and main source of all the later maps of the Mississippi and of the whole West of the United States."[108] Paullin and Wright note that, for the region of the Gulf of Mexico, it "long remained a model for cartographers."[109] Wheat calls it an "important cartographical monument" that represents "distinct advances in the mapping of the American West."[110] Wood deems it "the most important and influential map of the French period."[111] Martin and Martin say that it is "one of the great milestones in the cartographic history of North America in general and in Texas cartography in particular."[112]

Other authorities have not been content to regard Delisle's masterpiece with such unabashed–or, at least, such unqualified–reverence. As early as 1859, Thomassy suggested that Le Maire's "two other" 1716 maps, at that time unknown (one still is), probably were used by Delisle and de Fer for their 1718 maps.[113] Hamilton had this to say in 1934: "Comparing it [Le Maire's 1716 *Carte nouvelle*] with Delisle's print of 1718, one finds a great similarity in outline and legend. The perusal of the missionary's letters proves them to be the source for other features."[114] Delanglez went further, reminding us that Delisle's map is "clearly only a neat copy of these [Le Maire] maps in all that pertains to the geography of the Gulf coast."[115] The distinguished Jesuit scholar might have extended that observation well into the interior of Delisle's map, as others have recently done. Ehrenberg, for instance, notes that the entire region south of the Missouri "appears to have been taken from Father Le Maire's map."[116] Cumming, while saying that Delisle's map "makes a striking advance in its portrayal of the North American continent" and echoing Kohl's evaluation of it as a mother map, admits that Le Maire's

1716 map "or some copy of it" was used by Delisle in preparing his famous 1718 map.[117]

This map won for Guillaume Delisle the designation Premier Géographe du Roi, a title created especially for him.[118] But no one was more aware than Delisle himself of the debt that he owed Le Maire. Writing Bobé in 1718, prior to the publication of the map (which occured in June), Delisle thanked him for the latest collection of Louisiana material received. Noting that he used the "advice" of both Bobé and Le Maire, he states that he has rectified Le Maire's map in several places according to the memoirs of Iberville, Bienville, and Tonty "so that there are no spots on my map which have not been fixed by 1, 2, 3, or 4 degrees. It is a completely new map and, properly speaking, my own work. But the present state of the country and the location of the [Indian] Nations, several of which have moved since the time of M. d'Iberville, le Sueur, and others, are really known only by M. Le Maire. That is why I mainly used his memoirs on this situation"[119]

Launching into a discussion of Le Maire's names for the rivers of Texas, Delisle demonstrates a recognition of their value, as well as a reluctance to discard the names used on his 1703 *Carte du Mexique*. He solved the problem on his 1718 map by giving joint names–his and Le Maire's–to rivers judged to be the same, while retaining (erroneously) most of his other 1703 streams. Continuing, Delisle says:

> As I have many memoirs about these places, I can truly say that I have used the same memoirs on which Le Maire founded his ideas (for he could see but very few of the districts he describes). I could have done something more perfect with his map, which nonetheless is the most perfect until now. I know what we owe him on this subject, and I will not have this map engraved without acknowledging in the title that the Public is principally indebted to his remarks for our understanding of this country.[120]

Delisle kept his promise. Following the main title of his 1718 map appear the words "drawn after a great number of memoirs, amongst others, after those of M. Le Maire." That François Le Maire should be the only source cited by name indicates in fair measure the significance of his work to Delisle's accomplishments. Considering such a forthright admission on the face of the map, it is lamentable that more than two centuries had to pass before scholars started to recognize the extent of Le Maire's contribution to American cartography. But then, Delisle credited only Le Maire's memoirs, not his maps.

Le Maire did, however, have the satisfaction of hearing directly from the master cartographer and knowing that his own scholarly labors had not been in vain. On May 19, 1719–shortly before his return to France–Le Maire wrote Delisle, thanking him for his letter and expressing delight at the Louisiana maps received from him.[121] This letter confirms that Le Maire had also executed some "new" maps in 1718, on which were marked the routes of Saint-Denis and Derbanne.

This is fascinating information, because these maps have not been identified, only the 1716 map showing Saint-Denis's trip of 1714. Le Maire's 1717 maps, of course, could not have revealed the latest route taken, because the priest did not learn of it until Derbanne's return in November, when the journal of his trip was submitted. Yet, Delisle's 1718 map has both routes; so does the Vermale map usually accepted as 1717, although Cumming thinks that the date "is written in a different hand and with different ink." With his remark that the Delisle and Vermale maps indicate "a common source," Cumming nears the heart of the matter but fails to make the essential connection.[122] What we have now is evidence that Vermale's map shares its common features with the Delisle map because both ultimately came from the same source: François Le Maire's missing maps of 1718.

Sieur Vermale, as far as we can determine, never set foot on Louisiana soil. He styles himself a "Cornette de Dragons" in the title of his *Carte générale de la Louisiane ou du Miciscipi* (fig. 15).[123] He evidently was a draftsman, working in the ministry's depository, and it was his job to redraw rough maps received from the colonies.[124] Le Maire's 1716 *Carte nouvelle* had been reworked by someone in this manner. It is our belief that Vermale performed a similar function in 1718, neglecting to credit Le Maire, as the 1716 copyist had done, and that the 1717 date was added to the map later, as Cumming implies.

Le Maire, in his 1718 memoir, also mentions that he has sketched on his new maps the latest routes to the Spanish presidio: "The one which is northernmost is the last one, based on the journal of Sieur d'Erbanne."[125] As noted, the Vermale and Delisle maps show two routes to the Río Grande. In both cases the northern one is marked as the latest taken by Saint-

Denis. Because Le Maire was the primary cartographic source on this new venture, Vermale must have used one of the missionary's 1718 maps, forwarded early in the year, to produce his map of "1717."[126] As for Delisle's access to these 1718 maps, we have Le Maire's letter of May 19, 1719: "M. Bobé has probably transmitted to you my recent maps, in which you will find the last routes taken to go from Natchitoches to the Río Bravo."[127] How then, can we account for the appearance of this information on the Vermale and Delisle maps, except through the work of Le Maire?

Having carefully analyzed the Vermale and Delisle maps, we can only conclude that both must have issued from Le Maire prototypes. They are similar to each other and Le Maire not only in routes to Mexico, but also in the identification, placement, and spelling of rivers and Indian tribes. Several features shared by Vermale and Delisle, but not appearing on Le Maire's 1716 map, were derived from his memoirs, as Hamilton perceives. An example of this is the notation on the upper Missouri relating that Spaniards of New Mexico conducted an annual trade with the Indians of the region for "yellow iron"–gold.[128] Earlier, we saw that the characterization of all the Indians along the Gulf Coast as "nomadic cannibals" originates with Le Maire's 1716 map, seen in figure 10. Quite possibly, Le Maire added some of these elements to his maps of 1717 and 1718 as well.

Other features that cannot be attributed to Le Maire, but which are shared by Vermale and Delisle, leave us wondering who "borrowed" from whom between the two cartographers, but they do nothing to lessen Le Maire's profound influence. These similarities include the presence of "Etiopen errans" (roaming Ethiopians) in central Texas, a road that goes northwest from the Cenis and ends at a "Quiohouhahan" village, the word "Bagres" being added to Le Maire's Medina River, the appearance of "Bidaïe" Indians along the Trinity, and the survival of Delisle's 1703 "Sablonniere" River, to name a few. In substance, however, both maps clearly owe much to the missionary so maligned by his fellow Louisianians as a glutton and troublemaker, lax in his religious duties.[129]

There is yet another important map of the western country that we may link to Le Maire. It is the 1720 *Carte nouvelle de la partie de l'ouest de la province de la Louisiane* by Sieur de Beauvilliers, usually associated with the explorations of Jean-Baptiste

Bénard, Sieur de La Harpe (fig. 16).[130] This map, of which several copies are known,[131] is remarkable because of two Texas oddities: the road to Mexico is labeled "Route du Sieur d'Arbane," and the "Trupien" is shown as one of the main rivers draining into La Salle's bay. This river is the same described by Derbanne's journal as the "Irrupiens," because a tribe of Indians with that name lived on its banks.[132] That the entire Texas portion of the map was formed according to Derbanne's information, as presented on Le Maire's missing 1718 map(s), is suggested by the following statement the missionary made to Delisle in 1719: "This river of the Eripiames that you find added in the last map, as well as a few others [rivers], is rather a kind of torrent which was dry at the time of M. de St. Denis's first trips. *The best of these routes is the one I marked under the name of d'Erbanne* [emphasis added]. . . . The western corner of my map is so well-known nowdays, since d'Erbanne's trip, that you can trust what I last sent about it."[133] Interestingly, by the time Jean de Beaurain drew the other "La Harpe" map (fig. 17), Saint-Denis's name was once again on the route across Texas, but Derbanne's "Irrupiens," Le Maire's "Eripiames," and Beauvilliers's "Trupien" were perpetuated as the "Irrupien."[134] We suspect that missing sketch maps by Le Maire contributed to several other plans illustrating Beaurain's manuscript in the Library of Congress.

Regardless of his other accomplishments, Sieur de La Harpe was unable to provide much firsthand information about Texas, as he visited only the upper limits of the province. La Harpe's greatest contribution to the mapping of the Trans-Mississippi West came with his observations of the country above the Red River, not below it. His 1721 attempt to occupy San Bernardo Bay was an unqualified failure, as most scholars believe that he landed at Galveston Bay instead.[135] It seems almost certain that Beauvilliers, an engineer of France's Royal Academy of Sciences, produced his 1720 map working more from Le Maire–Derbanne than from La Harpe material.

Because so many of Le Maire's maps are lost or yet to be identified, it is still not possible to judge the full extent of his impact on state-of-the-art cartography in France and the world. As more of his maps become known, enabling us to discern how they were used by other mapmakers, perhaps his stature among those who helped chart the vast wilderness of North America will grow. Even so, we believe that the evidence presented here is ample to demonstrate a long-neglected truth: Delisle's "mother map" owes a considerable debt to the known maps of Le Maire. The evidence suggests that Delisle profited even more from the missionary's unknown maps, as did a host of mapmakers who followed their "collaboration." Had François Le Maire not sailed to Louisiana and applied his talents to geography, knowledge of the continent would not have come to the outside world as swiftly as it did, nor would cartography have developed in the methodical manner for which Guillaume Delisle has received most of the credit.

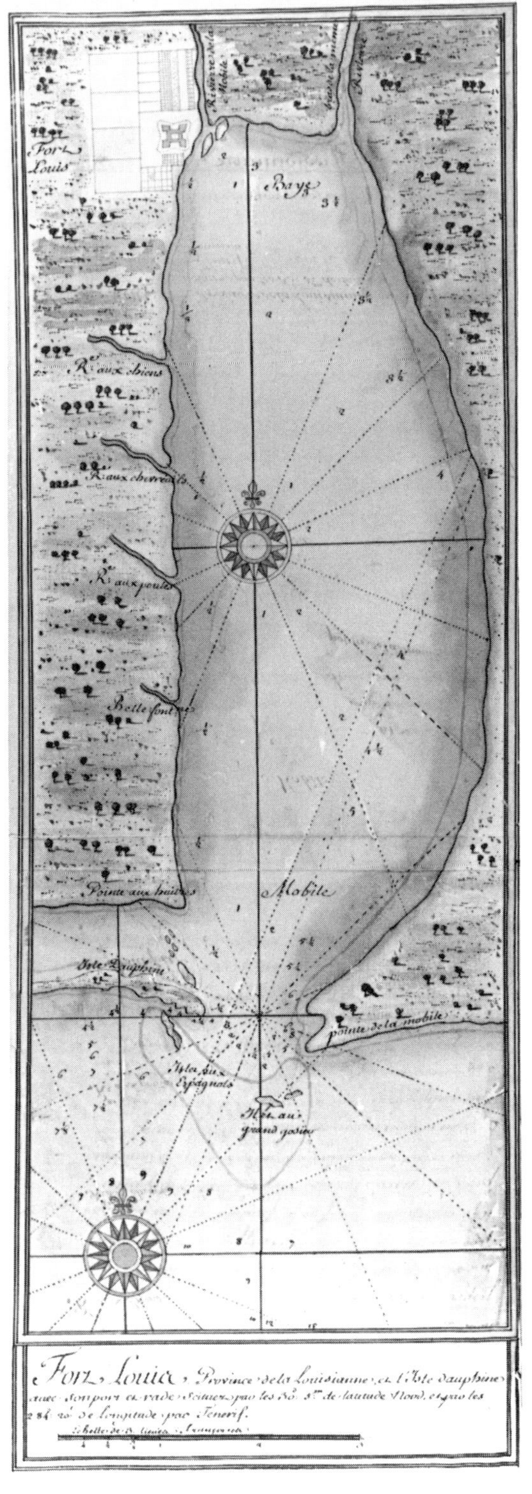

Fig. 5. Le Maire's map of Mobile Bay, made to accompany his letter of January 15, 1714, from Pensacola. Courtesy Newberry Library.

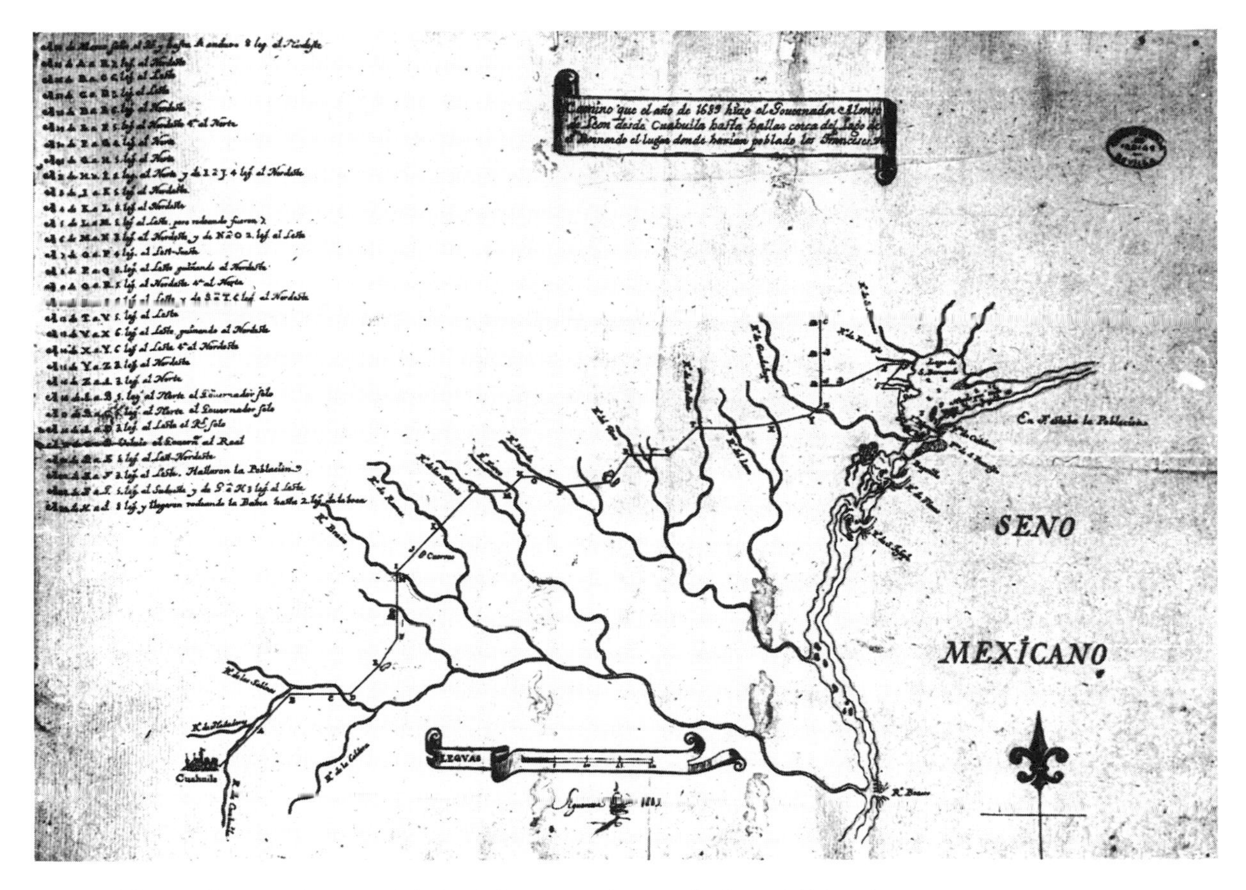

Fig. 6. Sigüenza's map of de León's route taken to find La Salle's colony in 1689. Original in AGI. Courtesy BTHC.

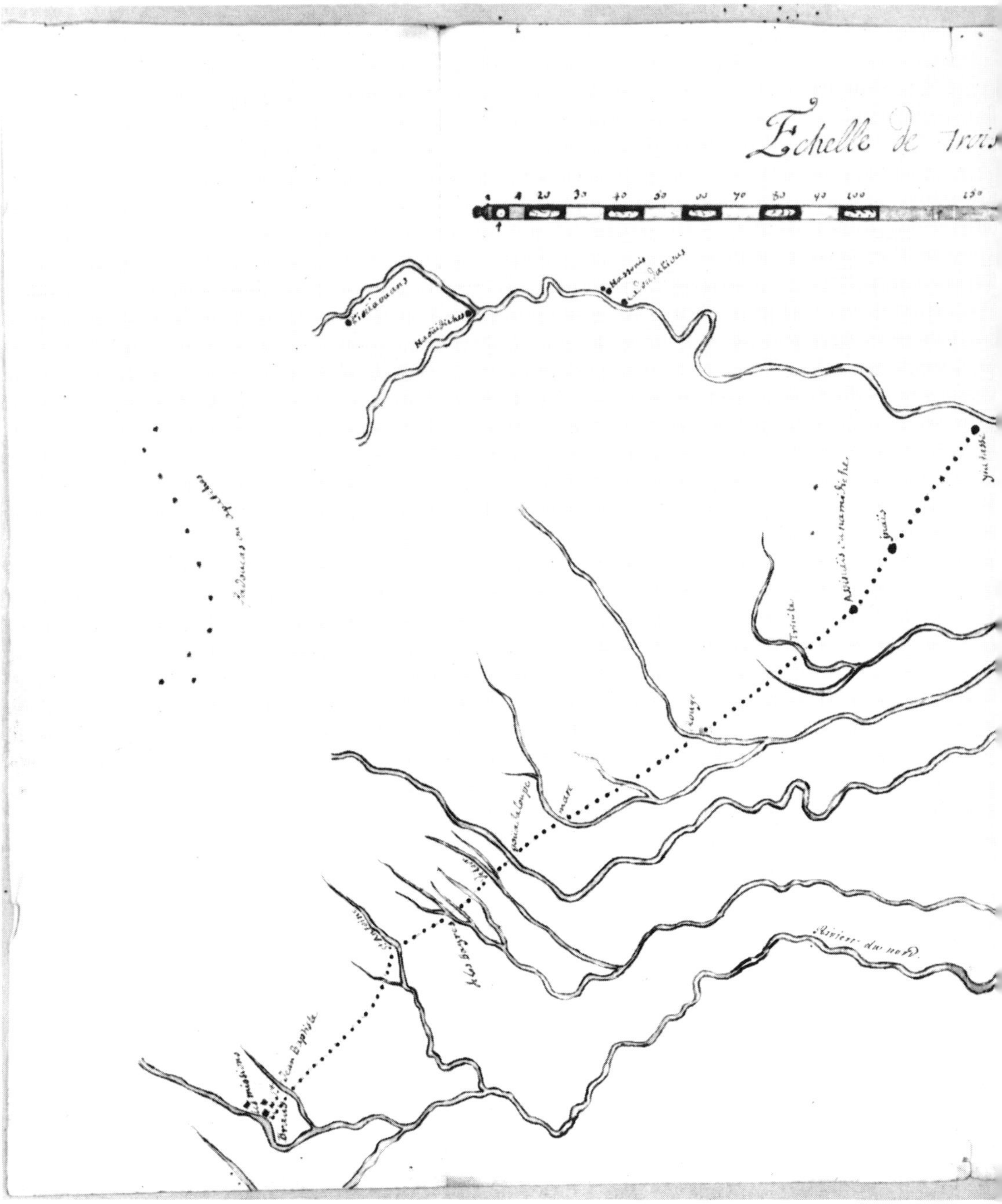

Echelle de trois

1 4 20 30 40 50 60 70 80 90 100 150

Trinité

Assinais

Riviere du nord

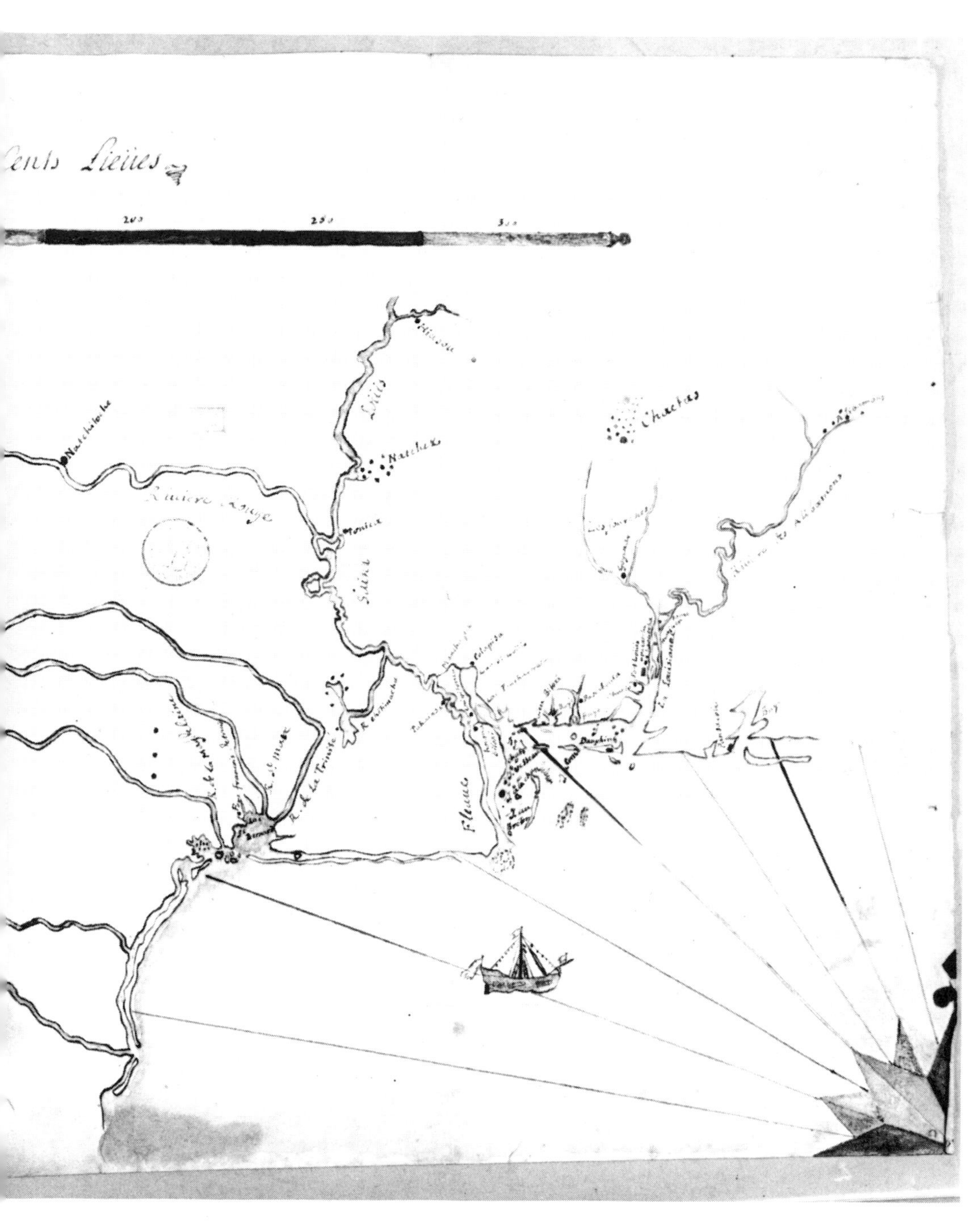

Fig. 7. Untitled map of Texas and Louisiana, said by Governor Cadillac to have been drawn by his son, 1716. Courtesy BN.

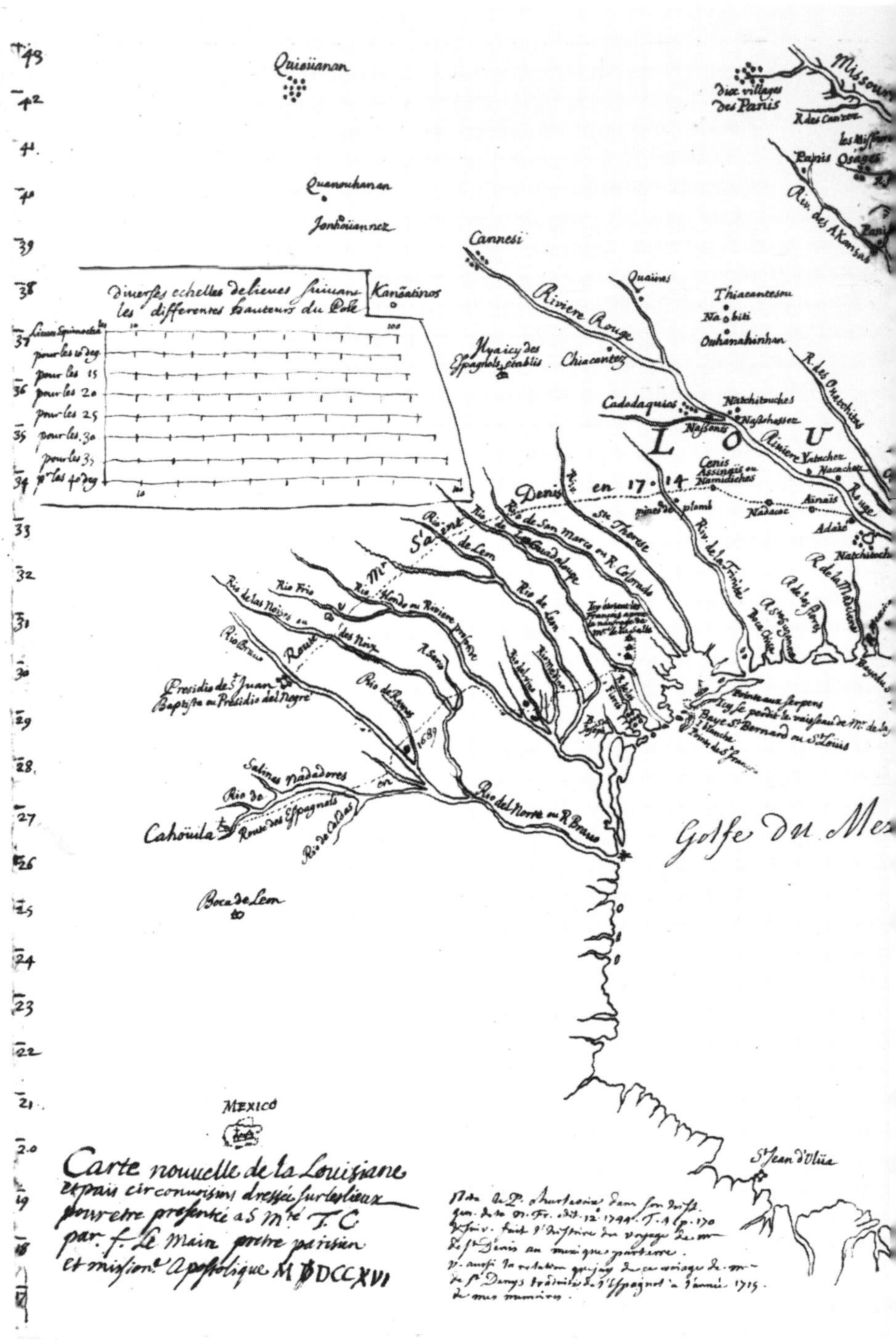
Quioüanan
Quanouhanan
Cannesi
Kanöatinos
Riviere Rouge
Quaires
Thiacantesou
Chiacantez
Cadodaquios
Natchitoches
Nassonis
L O U
Denis en 1714
Rio de San Marco ou R. Colorado
Rio de Leon
Rio Frio
Rio Bravo
Presidio de S.t Juan Baptista ou Presidio del Negro
Rio del Norte ou R. Brasso
Salinas Nadadores
Cahöuila
Boca de Leon
Golfe du Mex
MEXICO
S.t Jean d'Ulüa
Missouri
Riv. des AKansas
Rivere Rouge
diverses echelles de lieues suivant les differentes hauteurs du Pole
pour les 10 deg.
pour les 15
pour les 20
pour les 25
pour les 30
pour les 35
p.r les 40 deg
Carte nouvelle de la Louisiane et pais circonvoisins dressée sur les lieux pour etre presentée a S M.té T.C par f. Le Maire prestre parisien et mission.re apostolique MDCCXVI

Fig. 8. Le Maire's rough draft *Carte nouvelle de la Louisiane . . . 1716*, which Delisle used to produce his 1718 map. Original in SHM. Courtesy OSMRL.

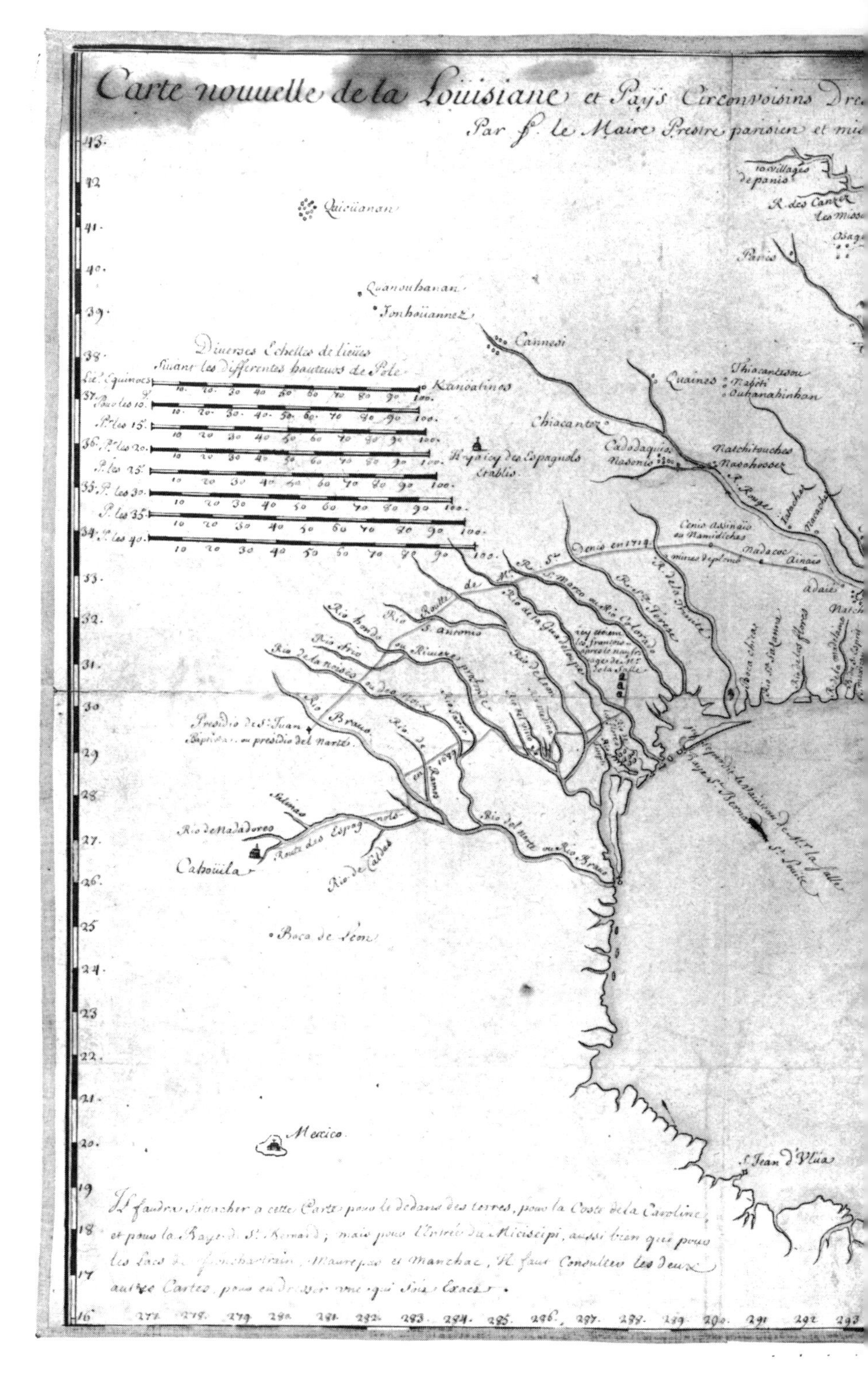
Carte nouuelle de la Louisiane et Pays Circonvoisins
Par f. le Maire Prestre parisien et
Diverses Echelles de lieües Suivant les differentes hauteurs de Pole
Kanoatinos
Cannesi
Quaines
Chiacantez
Cadodaquis
Nasonis
Natchitouches
R. Rouge
Nadacoc
Ainais
Adaies
Presidio de St Juan Baptista ou presidio del norte
Rio de Madadores
Cahoüila
Rio del norte ou Rio Bravo
Boca de Sem
Mexico
St Jean d'Vlua
Il faudra s'attacher a cette Carte pour le dedans des terres, pour la Coste de la Caroline, et pour la Baye de St Bernard; mais pour l'Entrée du Micisipi, aussi bien que pour les Lacs de Ponchartrain, Maurepas et Manchac, Il faut Consulter les deux autres Cartes, pour en dresser une qui Soit Exacte.

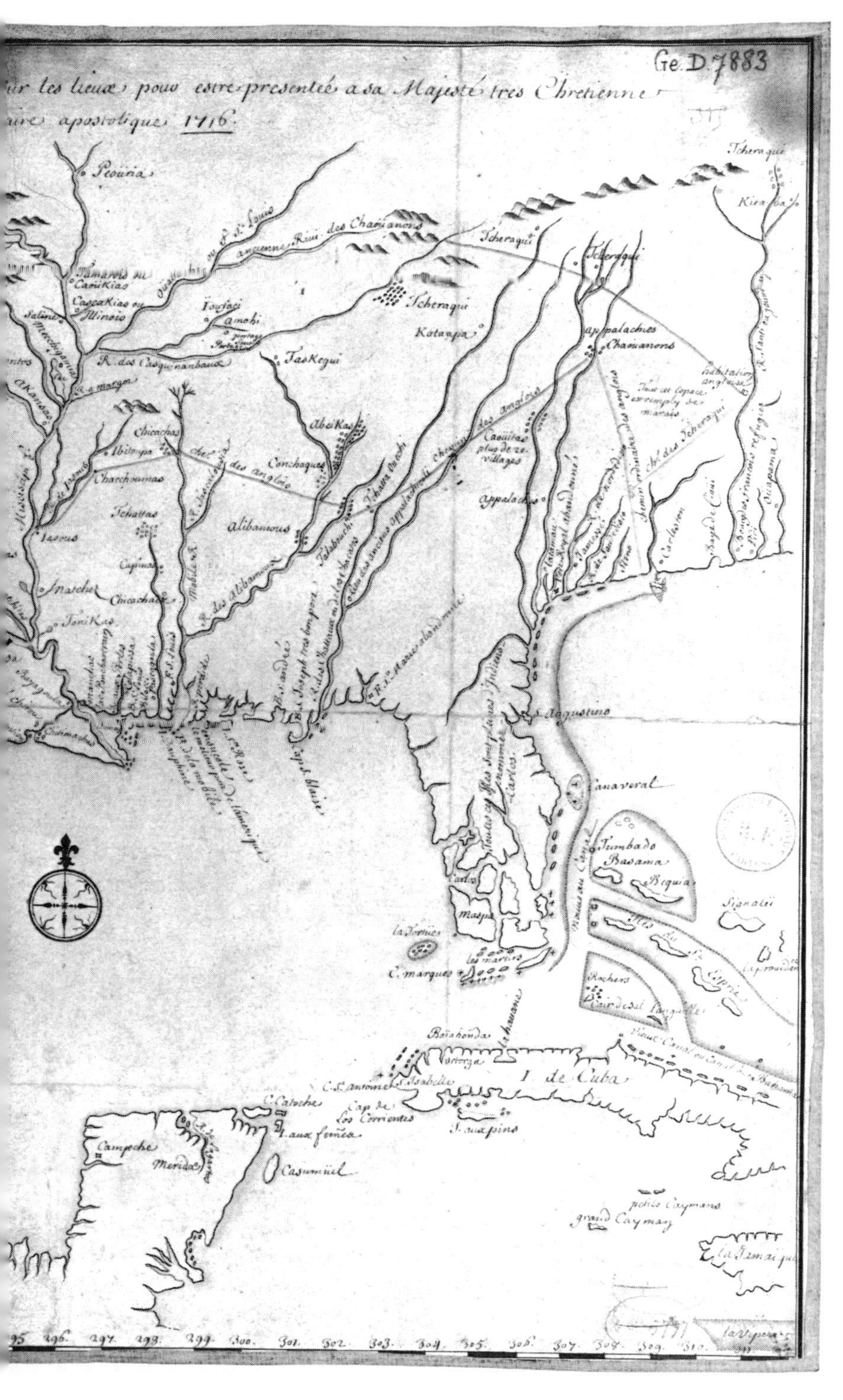

Fig. 9. A more polished copy of Le Maire's *Carte nouvelle,* drawn by an unknown hand. Courtesy BN.

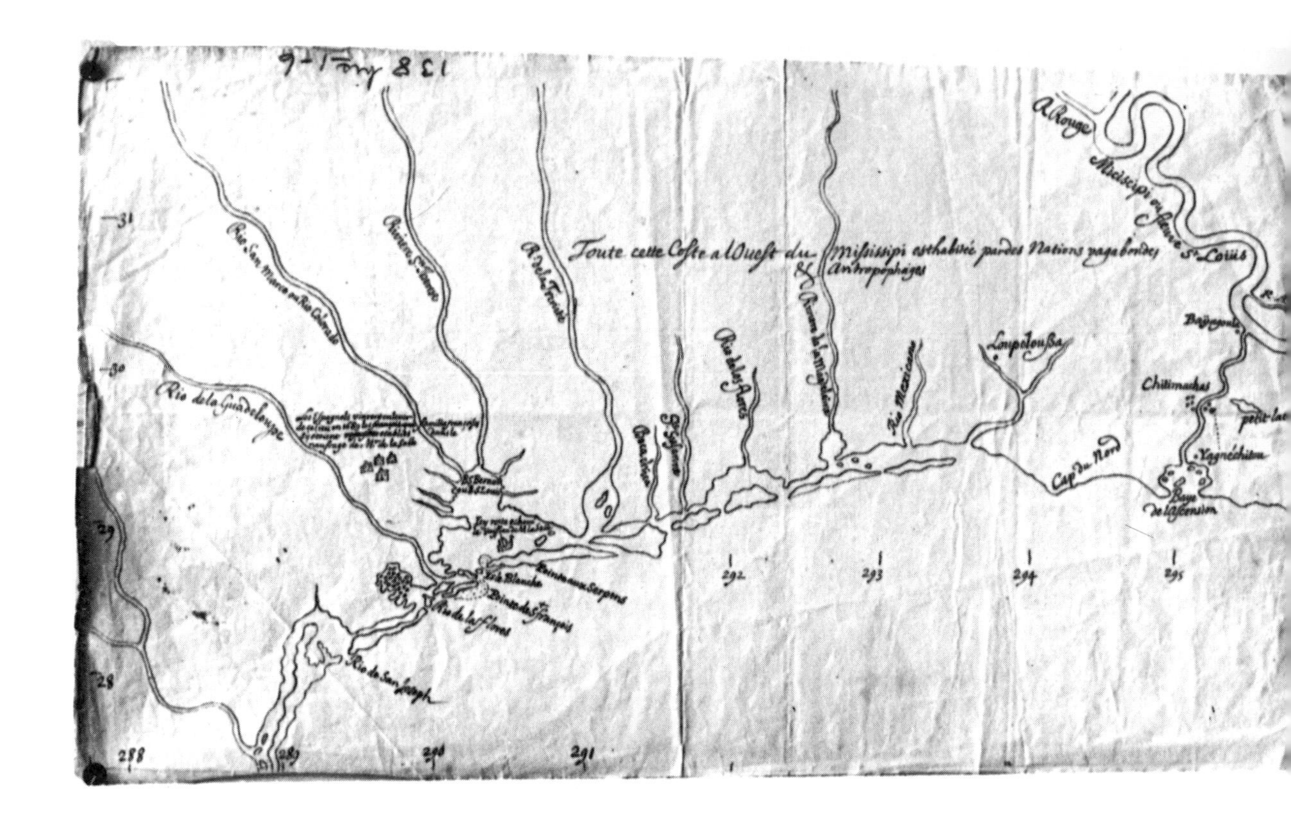

138 Tony 1-6
Toute cette Coste a l'Ouest du Mississipi est habitée par des Nations vagabondes & Antropophages
R. Rouge
Micisipi ou fleuve St. Louis
Rio San Marco ou Rio Colorado
Riviere St. Jacques
R. de la Trinité
Rio de la Guadeloupe
Rio de las Flores
Rio de San Joseph
Pointe aux Serpens
Pointe des François
Isle Blanche
Rio de las Auses
Rio Mexicano
Loupelousa
Cap du Nord
Bayagoula
Chitimachas
petit lac
Yagnéchitou
Baye de l'Ascension
31
30
29
28
288
289
290
291
292
293
294
295

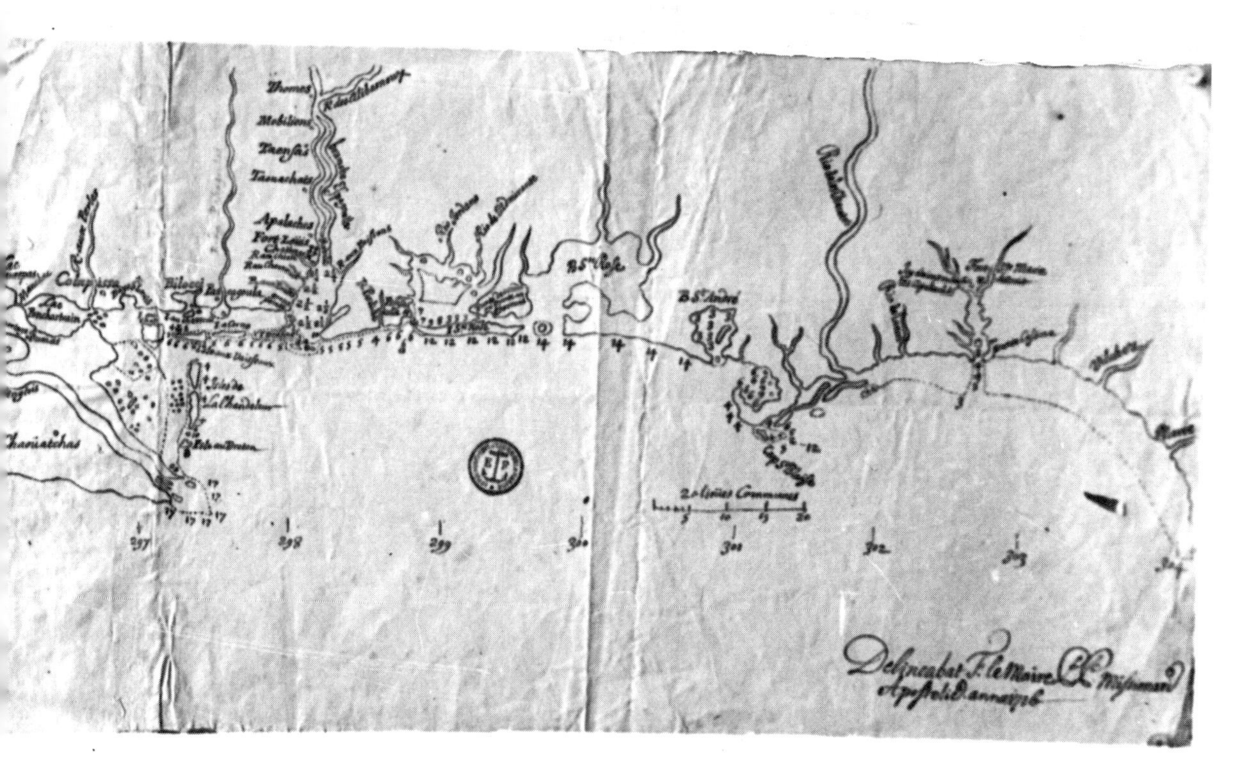

Fig. 10. Le Maire's untitled map of the northern Gulf Coast, 1716. Original in BN. Courtesy Newberry Library.

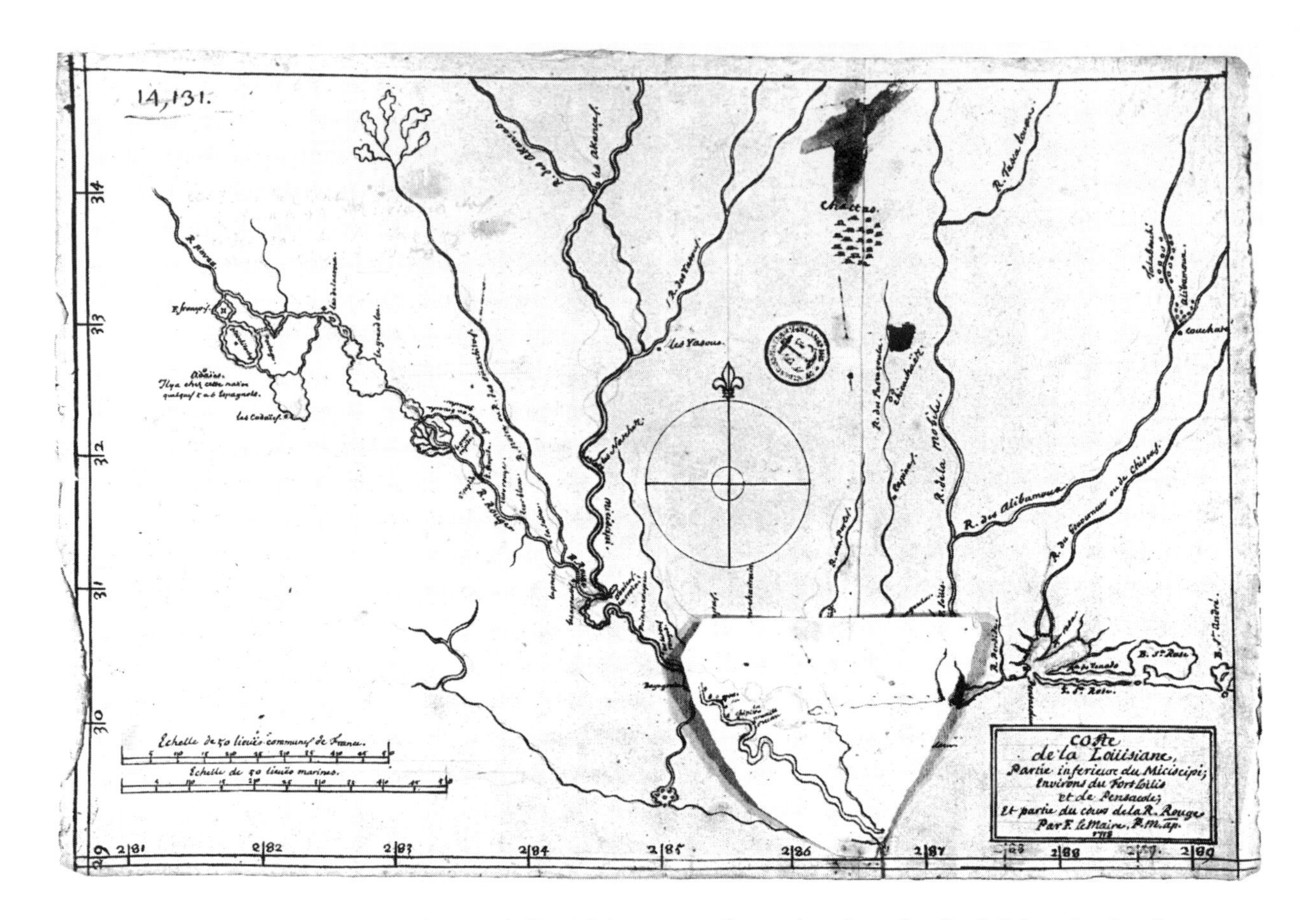

Fig. 11. A Le Maire map of 1718 showing the Red River in detail, found among Delisle's collection of sketch maps. Courtesy AN.

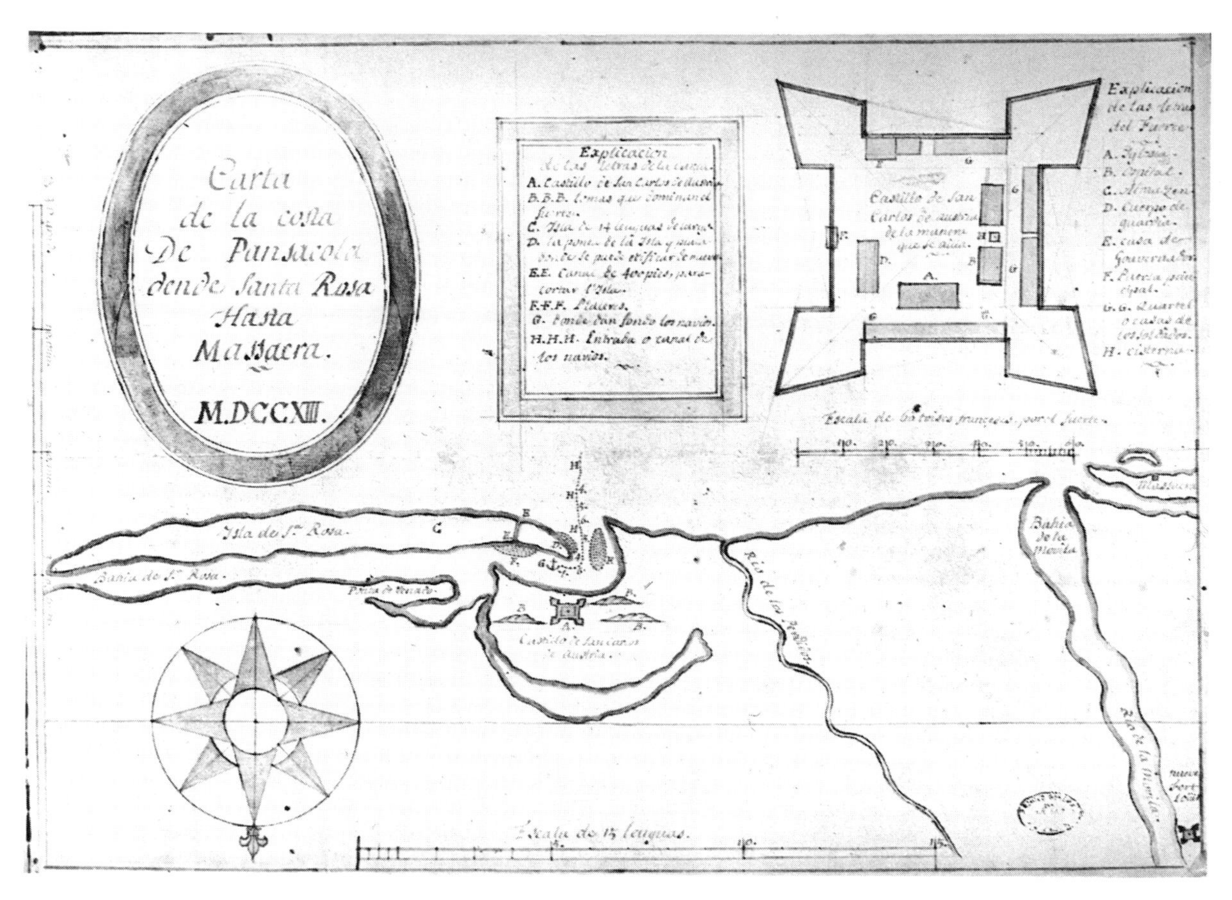

Fig. 12. Le Maire's 1713 sketch of the coast between Mobile Bay and Santa Rosa Sound, drawn while he was serving as a priest at Pensacola. Courtesy AGI.

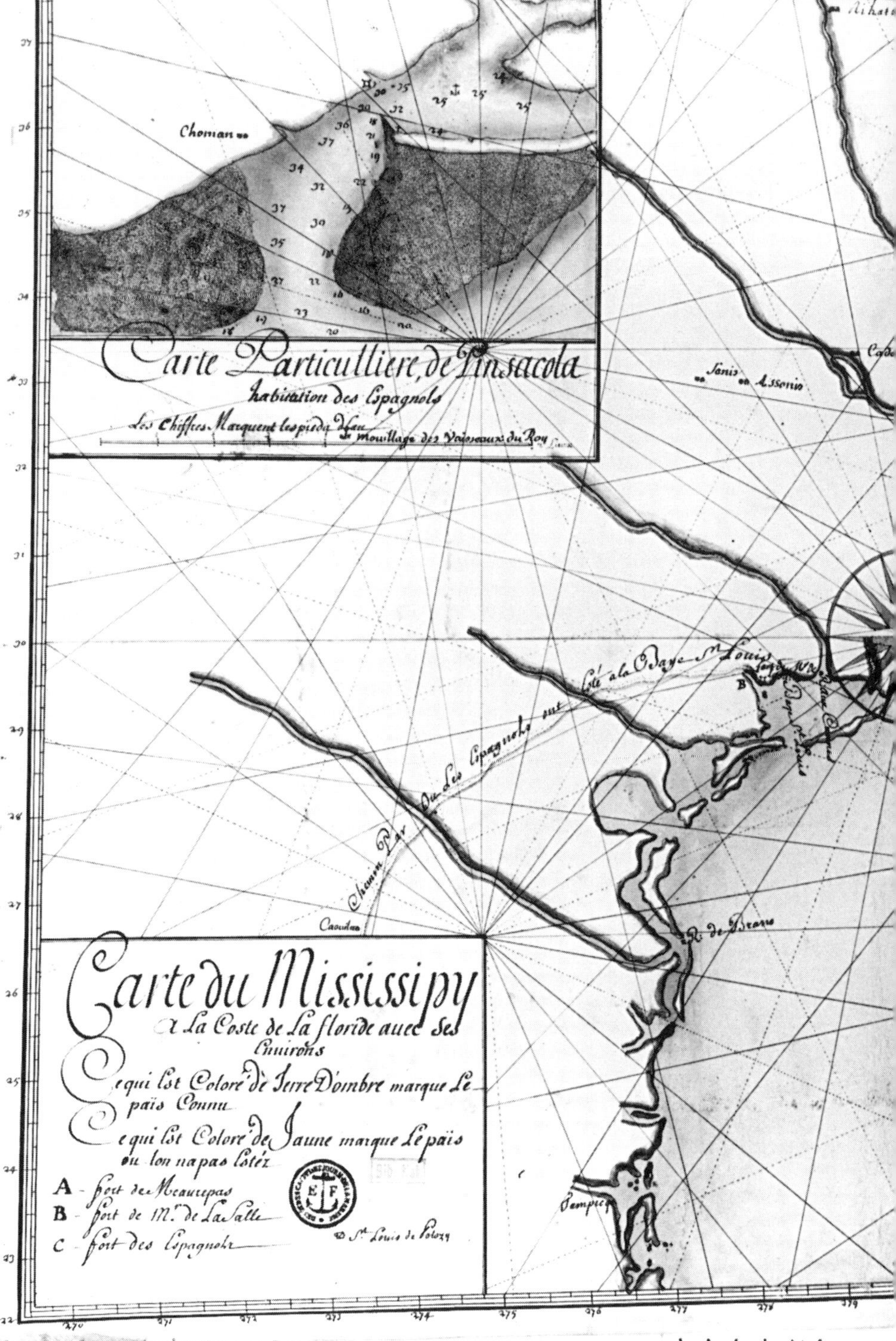
Choman
Carte Particulliere de Pinsacola
habitation des Espagnols
Les Chiffres Marquent les pieds d'Eau
Mouillage des Vaisseaux du Roy
Senis
Assonis
Carte du Mississipy
A La Coste de La floride auec Ses Enuirons
Ce qui Est Coloré de Terre D'ombre marque Le païs Connu
Ce qui Est Coloré de Jaune marque Le païs ou lon napas Esté
A - fort de Meaurepas
B - fort de Mr. de La Salle
C - fort des Espagnols
St. Louis de Potozy
Chemin Par Ou Les Espagnols ont Esté a la Baye St. Louis
R. de Bravo
Tampico

Fig. 13. An early French attempt, by an unknown hand, at charting the Gulf Coast, ca. 1700. Original in BN. Courtesy OSMRL.

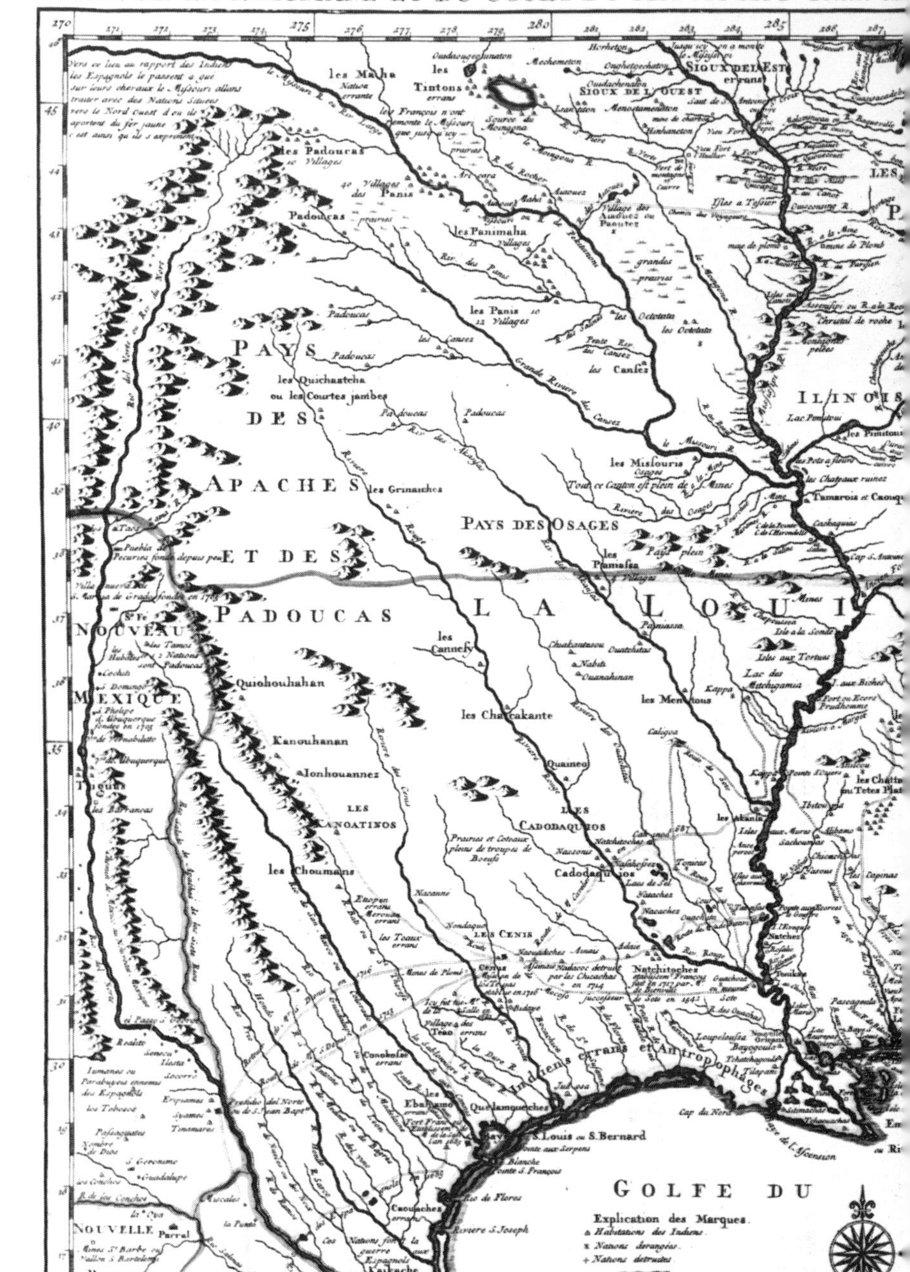
43
CARTE DE LA LOUISIANE ET DU COURS DU MISSISSIPI Dressée sur
PAYS DES APACHES ET DES PADOUCAS
NOUVEAU MEXIQUE
LA LOUI
PAYS DES OSAGES
ILINOIS
SIOUX DE L'EST
SIOUX DE L'OUEST
les Maha
les Tintons
les Padoucas
les Panimaha
les Panis
les Quichuatcha ou les Courtes jambes
les Grinatches
les Missouris
les Cannecy
les Chacakante
Quiohouhahan
Kanouhanan
Ionhouannez
LES KANOATINOS
les Choumans
LES CADODAQUIOS
Cadodaquios
Natchitoches
Natchez
LES CENIS
S. Louis ou S. Bernard
Quelamouches
Indiens errans et Antropophages
GOLFE DU
NOUVELLE BISCAYE
NOUVEAU ROYAUME DE LEON
Embouchure de Rio Bravo
Explication des Marques
Route de Soto et de Mocoso jusqu'au bas du Mississipi où ils furent poursuivis par les Indiens.

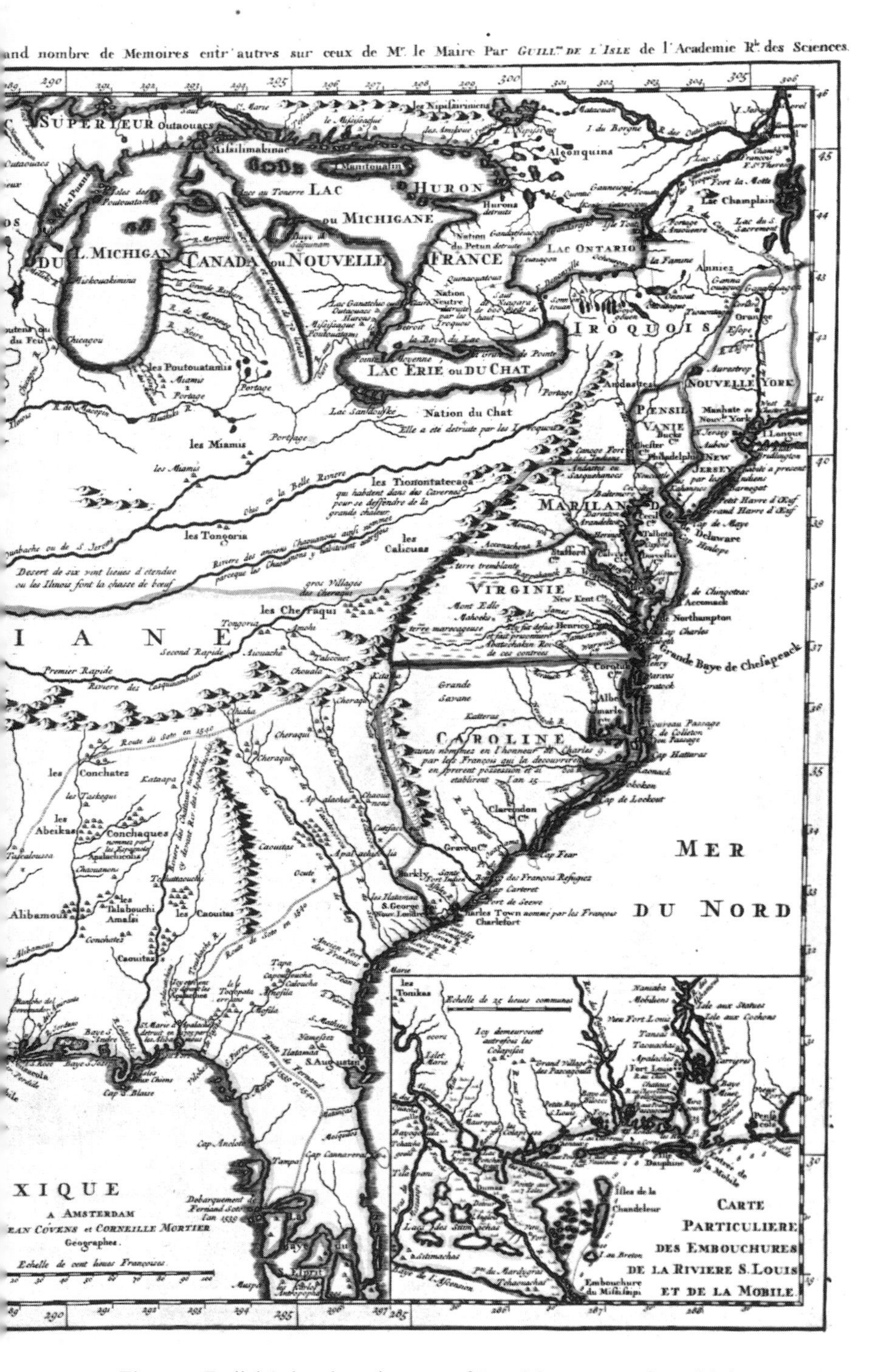

Fig. 14. Delisle's landmark map of Louisiana, 1718, for which Le Maire is credited. Courtesy BTHC.

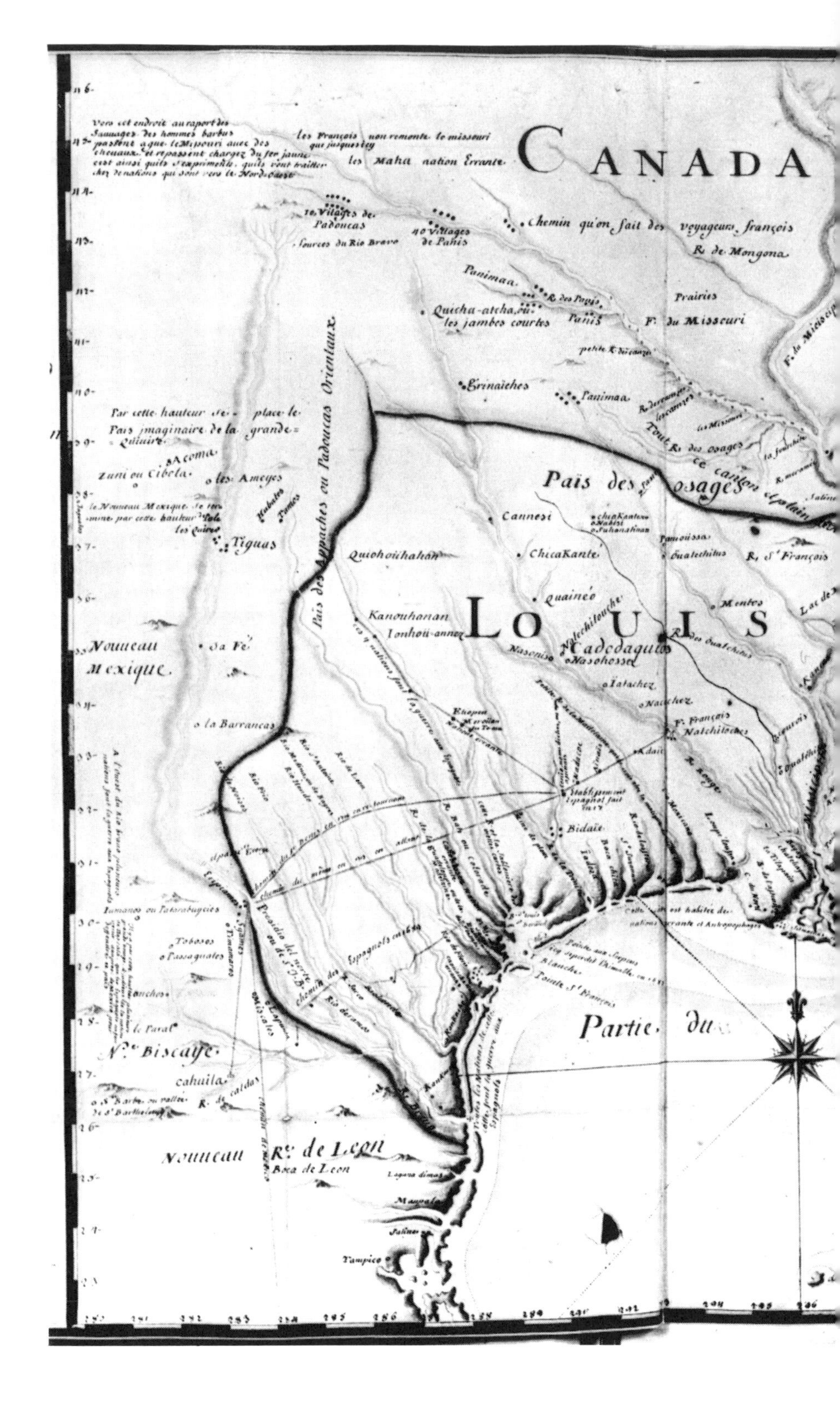
CANADA
les François non remonte le missouri qui jusques icy
les Maha nation Errante
10 Villages de Padoucas
Sources du Rio Bravo
40 Villages de Panis
Chemin qu'on fait des voyageurs françois
R. de Mongona
Panimaa
Quicha-atcha, ou les jambes courtes
Prairies
F. du Missouri
Erinaïches
Panimaa
Par cette hauteur se place le Pais imaginaire de la grande Quivira
Acoma
Zuni ou Cibola
les Ameyes
Tiguas
Païs des Appaches ou Padoucas Orientaux
Païs des osages
Cannesi
ChicaKante
Quiohoïchahan
Kanouhanan
Ionhoü-annez
Quainéo
LOUIS
Nasoniso
Cadodaquios
Nasohosse
Iatachez
Natchez
Mentos
R. des Ouatchitas
F. Françaïs Natchitoches
R. Rouge
Adaïs
Bidaïs
R. S.t François
Nouueau Mexique
Santa Fé
la Barrancas
Iumanos ou Patarabueies
Tobosos
Passaquates
Espagnols en 1689
Rio de la Sal
le Paral
N.e Biscaye
cahuila
Nouueau R.e de Leon
Boca de Leon
Tampico
Partie du

Fig. 15. Vermale's version of a Le Maire map (probably the one sent to France with his 1717 memoir) resembling in many aspects Delisle's 1718 map. Original in SHM. Courtesy Newberry Library.

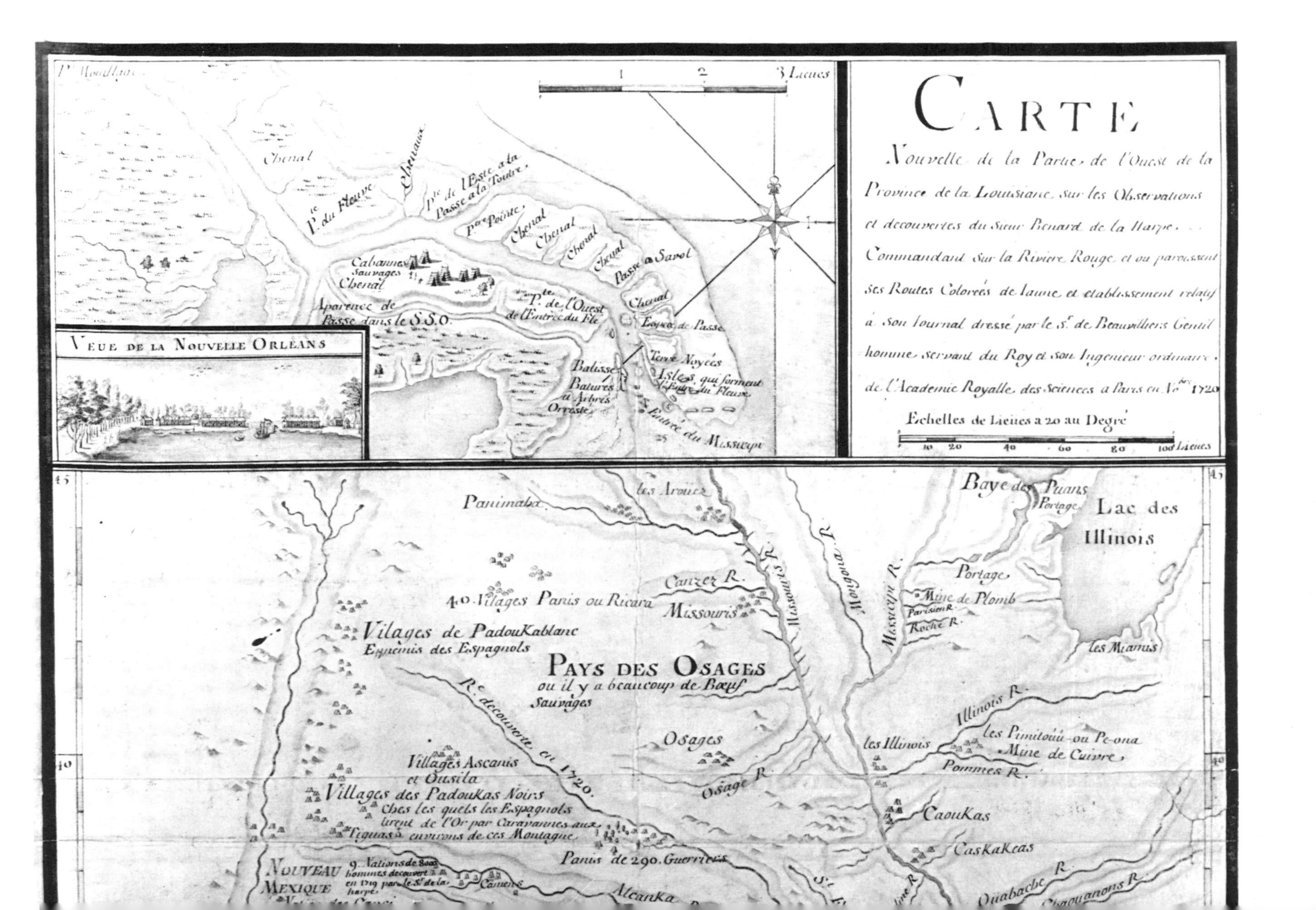
CARTE
Nouvelle de la Partie de l'Ouest de la Province de la Louisiane sur les Observations et decouvertes du Sieur Benard de la Harpe, Commandant sur la Riviere Rouge et ou paroissent ses Routes Colorées de Iaune et etablissement relatif à son Iournal dressé par le Sr. de Beauvilliers Gentil homme servant du Roy et son Ingenieur ordinaire de l'Academie Royalle des Sciences a Paris en Nbre. 1720
Echelles de Lieues a 20 au Degré
Veue de la Nouvelle Orleans
Chenal
P. du Fleuve
Passe a Savol
Aparence de Passe dans le S.S.O.
P. de l'Ouest de l'Entrée du Fle
Balisse
Entrée du Missisipi
Baye des Puans
Lac des Illinois
Pays des Osages
ou il y a beaucoup de Boeufs Sauvages
Vilages de PadouKablanc
Ennemis des Espagnols
40 Vilages Panis ou Ricara
R. decouverte en 1720.
Villages Ascanis et Ousila
Villages des Padoukas Noirs
Panis de 290. Guerriers
Nouveau Mexique
Missouris R.
Missisipi R.
Mine de Plomb
Illinois R.
Mine de Cuivre
Osage R.
Ouabache R.
CasKaKeas
CaouKas

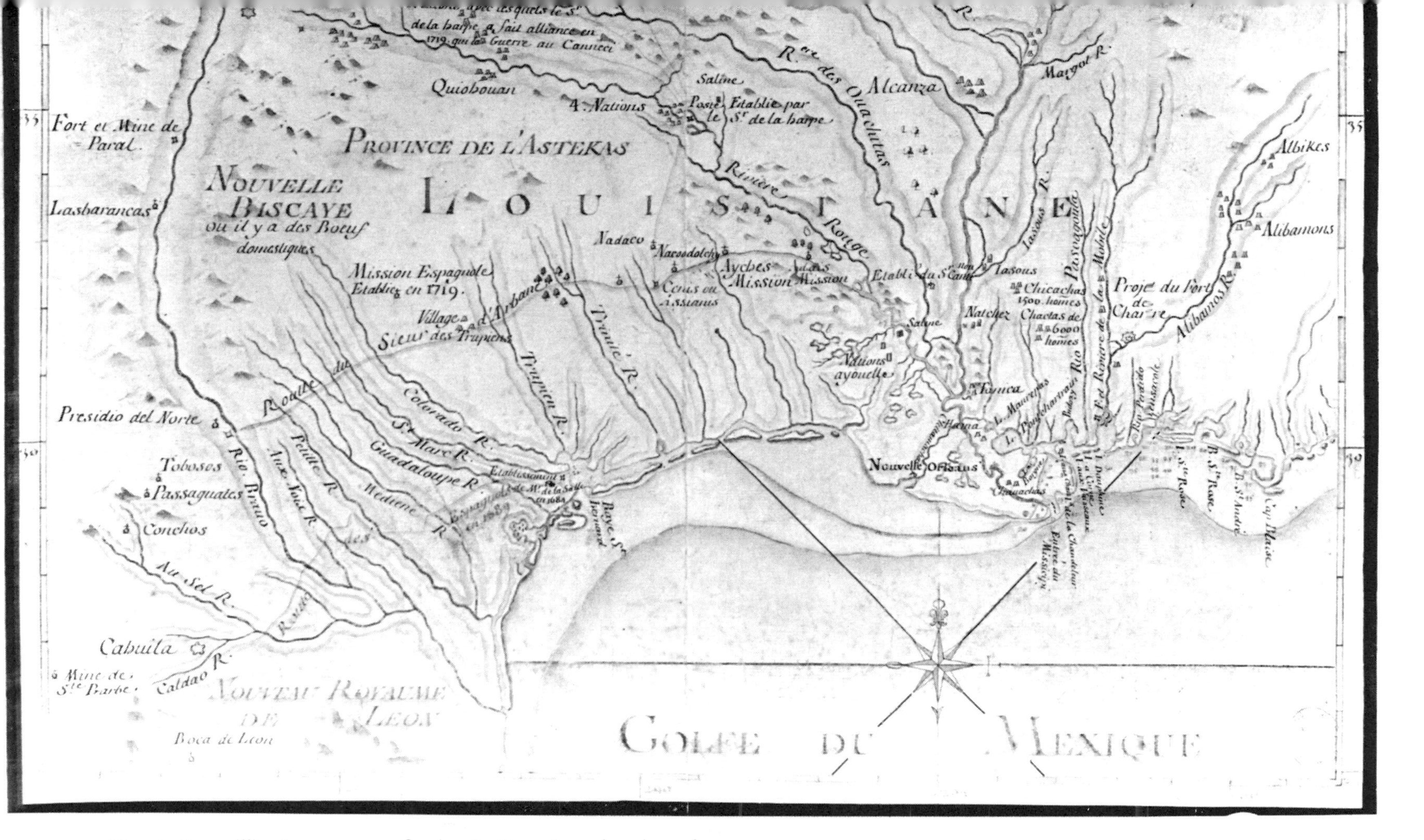

Fig. 16. Beauvilliers's 1720 map reflecting La Harpe's explorations above the Red River but relying on Le Maire and his sources for other features. Original in BN. Courtesy OSMRL.

Fig. 17. Detail from the "western part of Louisiana," usually attributed to Beaurain, ca. 1725. Courtesy LC.

Abbreviations Used in Notes

AC	Archives des Colonies (a record group now at the AN)
AGI	Archivo General de Indias, Seville
AGN	Archivo General de la Nación, Mexico
AN	Archives Nationales, Paris
ASH	Archives du Service Hydrographique de la Marine (refers to materials now in the AN)
BN	Bibliothèque Nationale, Paris
BSH	Bibliothèque du Service Hydrographique de la Marine (a designator for atlases now at the SHM)
BTHC	Barker Texas History Center, University of Texas, Austin
LC	Library of Congress, Washington, D.C.
MPA	*Mississippi Provincial Archives*
OSMRL	Old Spanish Missions Research Library, Our Lady of the Lake University, San Antonio
SHM	Service Historique de la Marine, Vincennes.

Notes

Oliván Rebolledo–Saint Denis Maps

1. Alvarez de Pineda's map, from AGI, Mapas y Planos, is reproduced in James C. Martin and Robert Sidney Martin, *Contours of Discovery: Printed Maps Delineating the Texas and Southwestern Chapters in the Cartographic History of North America, 1513–1930*, p. 11.

2. Basic studies include William Edward Dunn, *Spanish and French Rivalry in the Gulf Region of the United States, 1678–1702: The Beginnings of Texas and Pensacola;* Dunn, "The Spanish Search for La Salle's Colony on the Bay of Espíritu Santo, 1685–1689," *Southwestern Historical Quarterly* 19, no. 4 (Apr., 1916): 323–69; Herbert Eugene Bolton, "The Location of La Salle's Colony on the Gulf of Mexico," *Southwestern Historical Quarterly* 27, no. 3 (Jan., 1924): 171–89; and Robert S. Weddle, *Wilderness Manhunt: The Spanish Search for La Salle*.

3. Enríquez Barroto's diary is translated by Robert S. Weddle in *La Salle, the Mississippi, and the Gulf: Three Primary Documents*, ed. Robert S. Weddle, pp. 149–205. Barroto's map survives, to an extent, in several maps drawn by a Flemish-Spanish pilot, Juan Bisente, who was based at Havana and serving in the Armada de Barlovento. These maps, dated 1696 and 1700, are now in the BN and are the subject of an investigation by the authors.

4. Cárdenas's *Planta cosmográphica del Lago de San Bernardo*, AGI, Mapas y Planos, is described in Woodbury Lowery, *The Lowery Collection: A Descriptive List of Maps of the Spanish Possessions within the Present Limits of the United States, 1502–1820*, item 200. With reference to Lowery, it should be noted that since his compilation the maps in the AGI have been separated from their original *legajos* and placed in a separate section, Mapas y Planos. The Cárdenas map is reproduced (from a hand-drawn copy at the BTHC) in Weddle, *Wilderness Manhunt*, plate 11.

5. Sigüenza's *Mapa del camino que el año de 1689 hizo el Gobernador Alonso de León*, described in Lowery, *Lowery Collection*, item 193, is reproduced in Weddle, *Wilderness Manhunt*, plate 8, from the BTHC's hand-drawn copy of the AGI original. For the original, see fig. 6 in Jackson and De Ville, "Le Maire and the 'Mother Map' of Delisle," this volume.

6. *Mapa del Viaxe que el año 1690 hizo el Gobernador Alonso de León desde Cuahuila hasta la Carolina*, AGI, Mapas y Planos, is reproduced in James P. Bryan and Walter K. Hanak, *Texas in Maps*, plate 7, and described in Lowery, *Lowery Collection*, item 197.

7. Terán's *Mapa de la provincia donde habita la nación Casdudacho,* AGI, Mapas y Planos, is reproduced in Herbert Eugene Bolton, *Texas in the Middle Eighteenth Century: Studies in Spanish Colonial History and Administration,* frontispiece, and described in Lowery, *Lowery Collection,* item 199.

8. Jean-Baptiste Bénard, Sieur de La Harpe, *The Historical Journal of the Establishment of the French in Louisiana,* ed. Glenn R. Conrad, p. 35. The French occupation of the Gulf Coast is treated in Marcel Giraud, *A History of French Louisiana: The Reign of Louis XIV, 1698–1715,* trans. Joseph C. Lambert; and Jay Higginbotham, *Old Mobile: Fort Louis de la Louisiane, 1702–1711.*

9. Charmion Clair Shelby, "International Rivalry in Northeastern New Spain, 1700–1725" (Ph.D. diss., University of Texas, 1935), pp. 111–13.

10. Ibid., pp. 105–10.

11. Cadillac to Pontchartrain, Oct. 26, 1713, AC, C13A, 3:1–93, LC transcript. The LC has transcripts of this entire record group. The BTHC has a five-volume set of transcripts (cataloged as nos. 735–39) from the AN.

12. Carlos E. Castañeda, *Our Catholic Heritage in Texas,* 2:18.

13. Saint-Denis's first declaration, June 22, 1715; Spanish copies in Archivo San Francisco el Grande, Mexico, 8 (photostat); AGN, PI 181:3–9, and AGN, H 27:121 (transcripts); French copy in AC, C13A, 4:1010 (transcript). All in BTHC transcripts.

14. Cadillac to Pontchartrain, Sept. 18, 1714, AC, C13A, 3:531, LC transcript. The role of the Talon brothers in western exploration is reviewed by Weddle, "The Talon Interrogations: A Rare Perspective," in his *La Salle,* pp. 220–24.

15. Salinas Varona to the viceroy, Aug. 29, 1713, AGI, Mexico 61-1-34 (old number), BTHC transcript.

16. Moro to Gallut, Sept. 20, 1715, AC, C11A, 36:415; Moro to Crozat, July 18, 1716, AC C11A, 36:418, LC transcripts. Concerning Moro, see Shelby, "International Rivalry," pp. 130, 131, 136.

17. Saint-Denis's first declaration.

18. The mapmaking role of Moro and the unidentified Englishman emerged only when Saint-Denis gave his second declaration, Sept. 18, 1717, in "Testimonio de diligencias," AGI, Mexico 61-6-35 (old number), pp. 34–59, BTHC transcript; translated in Charmion Clair Shelby, "St. Denis's Declaration concerning Texas in 1717," *Southwestern Historical Quarterly* 26, no. 3 (Jan., 1923): 165–83.

19. Joseph Antonio de Espinosa, "Dictamen," Aug. 15, 1715, AGN, PI 181:10–15, BTHC transcript. See also Espinosa's opinion of Nov. 30, 1716, and the more detailed plan of *fiscal* Velasco, Nov. 30, 1716, AGN, PI 181:134–38, 139–80, BTHC transcript. All of these opinions mention Saint-Denis's map and the threat it implied.

20. Report of Junta General, Aug. 22, 1715, AGN, PI 181:16.

21. Ibid. Henry R. Wagner, in *The Spanish Southwest, 1542–1794: An Annotated Bibliography,* 2:333, cites one of Oliván's letters in the LC stating that he had been "nominated to make an examination of Texas and Louisiana" by 1716.

22. There has been much confusion as to the identity of Saint-Denis's bride, as no record of their marriage has been found, nor is there sufficient documentation as to the names of her parents. The best information seems to be that she was the daughter of Captain Ramón's daughter—accounting for her surname Sánchez, or Sánchez Navarro—and not of Diego the younger

as most writers have claimed. Some interesting notes on the question are found in folder 3 of Charmion Clair Shelby's papers, BTHC, box 3C141.

23. Shelby, "International Rivalry," p. 141.

24. For detailed treatment see Shelby, "St. Denis's Second Expedition from Louisiana to the Río Grande, 1716–1719" (Master's thesis, University of Texas, 1927) and her preliminary article, "St. Denis's Second Expedition to the Río Grande," *Southwestern Historical Quarterly* 27, no. 3 (Jan., 1924): 190–216.

25. Salinas to the king, Jan. 20, 1717, and Salinas to the viceroy, Feb. 15, 1717, AGI, Mexico 61-6-35. See also the series of Salinas's letters on how to protect New Spain from the French in AGI, Guadalajara 67-3-28 (old numbers), BTHC transcripts.

26. Minutes of the Council of Marine, Aug. 29 and Oct. 9, 1716, AC, C13A, 4:205, 389–94, LC transcripts.

27. Alarcón to the viceroy, June 27, 1717, AGI, Mexico 61-6-35 (old number), BTHC transcript.

28. Oliván, "Informe," Dec. 24, 1717, AGI, Mexico 61-6-35, BTHC transcript, translated in Shelby, "St. Denis's Second Expedition," pp. 75–100. This important document is also found in translation in the Bolton Collection, Bancroft Library, University of California, Berkeley. It is difficult to establish when Oliván began his various studies on the French intrusion, but he was named to investigate the problem–and Saint-Denis–by decree of July 20, 1717. He had already prepared a report for the king, over Viceroy Linares's signature, that accompanied the Junta's report of Aug. 22, 1715. Sometime after April 24, 1717, Oliván gave Viceroy Valero more information on the attempts to safeguard Texas (undated BTHC transcript, box 2Q246; copy also in Bolton Collection). Another report to the king, July 28, 1717–only a week after his latest commission–is summarized in AGI, Guadalajara 67-3-28, BTHC transcript. Thus it is evident that Oliván's interest in Texas and the northern frontier was well developed by the time he prepared his comprehensive "Informe" of Dec. 24, 1717.

29. Oliván, "Informe."

30. Saint-Denis's second declaration. Oliván's examination of the prisoner began on Sept. 1.

31. Reference is to Oliván's report of Dec. 24, 1717, and the map that he made for it, described in note 39 below.

32. Oliván, "Parecer," Nov. 4, 1717, AGI, Mexico 61-6-35, BTHC transcript.

33. Oliván, "Informe."

34. These maps are described in a series of AGN publications, *Catálogo de Ilustraciones*. The catalog assigns to them nos. 197.1–197.7.

35. The map that forms the basis for the designated maker and date of the set is AGN no. 197.5.

36. AGN nos. 197.5 and 197.1. The latter is badly water damaged but contains interesting information on Coahuila and Texas, extending eastward to the Río de San Marcos.

37. AGN nos. 197.7 (central Mexico) and 197.4 (coastline).

38. AGN nos. 197.2, 197.3, and 197.6. These are our figs. 3, 1, and 2, respectively.

39. Fig. 4 bears the legend, "Mapa Geográfico que presentó con su Ynforme al Virrey de la Nueva España, Don Juan de Oliván Rebolledo, oydor de la Real Audiencia de Mexico, en 18 de Diciembre de 1717; y á consequencia del reconocim^{to} que hizo de estas Provincias de Orden de S.E.," AGI, Mapas

y Planos, and is reproduced from a copy in the Texas State Archives. This Oliván map is Lowery's item 280 and has often been reproduced, though not heretofore linked to its companion pieces. It is exactly the same size as the others.

40. Oliván, "Informe." The notes at the bottom of figs. 1 and 2 serve as titles. The note on fig. 4 reads, "En este mapa reconocerá VE con mas extensión, lo que expreso en el Ynforme de 18 [*sic*] de Diz[c] de [1]717, a que acompaña." Each note is signed with the rubric "JOR."

41. The note on fig. 1 (AGN 197.3) reads, "Con este mapa podra VE hacer juicio en que forma esta confinante la Canada y N[a]. Inglaterra con la Florida y Luysiana – esta con los Asinais, Coaguila, y León, y estos con nuebo MEXICO: no lo he podido deducir mas exacto de las noticias – adquira en Francia no ha salido mapa de los Lignes confinantes a la Luysiana por el poniente." This map evidently is the one Oliván prepared so that Viceroy Valero could see "the distance existing between our countries and those occupied by the French" (undated Oliván typescript, BTHC; see note 28 above).

42. Saint-Denis's second declaration.

43. Fig. 2 (AGN 197.6) bears the inscription, "He formado este mapa mas extenso de el que pase en manos de VE para pueda VE reconocer mejor los confinantes de la N[a]. Francia con el Reyno de León, y Tejas; y los Ríos Intermedios; y sus desembocaduras al Seno Mexicano." It is the "more detailed" map given to Valero sometime before April, 1717. Oliván commented that the two maps were "not as accurate as I would want them to be, due to the poor information we are able to secure," and admitted that knowledge was limited "about the lands that the foreign machination wants to take away from us." Nonetheless, Spain could prevent the loss, "if it is so resolved" (undated typescript, BTHC). Oliván's reports and maps were dedicated to such a resolution.

44. Reference is to Louis Hennepin's *Carte de la Nouvelle France et de la Louisiane nouvellement découverte,* published in Paris in 1683, and I. Rouillard's *Carte généralle de la Nouvelle France où est compris la Louisiane gaspésie et le Nouveau Mexique,* published in Paris in Chrétien Le Clercq's *Premier établissement de la foi* (1691).

45. Saint-Denis's second declaration.

46. Fig. 3 (AGN 197.2), unlike figs. 1, 2, and 4, bears no inscription. It seems to be Oliván's working copy for fig 4, which the viceroy forwarded to Spain with related documents. Thus, fig. 4 wound up in the AGI, while the other two maps in the series (figs. 1 and 2) and Oliván's rough draft (fig. 3) remained in Mexico, becoming part of the AGN.

47. Oliván, "Informe."

48. Ibid.

49. Ibid. El Cuartelejo, an Apache settlement, has been placed at the junction of Mustang and Adobe creeks in present Kiowa County, Colorado (José Antonio Pichardo, *Pichardo's Treatise on the Limits of Louisiana and Texas,* 1:35n).

50. Studies on Gálvez's reforms include Luis Navarro García, *Don José de Gálvez y la comandancia general de las provincias internas del norte de Nueva España,* and Herbert I. Priestley, *José de Gálvez, Visitor-General of New Spain.*

51. Wagner, *Spanish Southwest,* 2:333, 334. Wagner's document assigns boundaries to the province of "Nuevas Philipinas," perhaps the earliest definition of Spanish Texas. The province was to be bounded on the east by the Mississippi River; on the south by the coast; on the west by the Medina River,

from its source to its mouth in the Gulf; on the north by an imaginary line extending from the Medina's source to the point at which the "Colorado of the Cadodachas" (Red River) joined the Mississippi.

52. Alfred Barnaby Thomas, ed. and trans., *After Coronado: Spanish Exploration Northeast of New Mexico, 1696–1727*, pp. 39–40.

53. Oliván, "Parecer," Oct. 2, 1724, AGI, Mexico 62-1-41, BTHC transcript.

54. One of the best studies of Barreiro's and Oliván's contributions to the Rivera inspection is Henrietta Murphy, "Spanish Presidial Administration as Exemplified by the Inspection of Pedro de Rivera, 1724–1728" (Ph.D. diss., University of Texas, 1938), pp. 175–77. Although Barreiro did not draw an individual map of Texas, he included the province in his map of the entire inspection (no. 6), missing from the set in the AGI, Mapas y Planos. The general map survives through a copy made by Luis de Surville in 1770, now in the British Library and reproduced as item 115 in Carl I. Wheat, *Mapping the Transmississippi West, 1540–1861*, vol. 1, opp. p. 82. Comparison of the eastern portion with figs. 3 and 4 reveals the way in which Oliván's concept of the northern frontier influenced Barreiro, especially in the east-west elongation of Texas. Barreiro's "Descripzión," Feb. 10, 1730, which accompanied the maps, is in AGI, Mexico 61-2-12, BTHC transcript; copy also in Bolton Collection.

55. Reproduced as fig. 8 in Jackson and De Ville, "Le Maire and the 'Mother Map' of Delisle," this volume.

56. Salinas to the viceroy, Feb. 15, 1717, AGI, Mexico 61-6-35, BTHC transcript. Attached was an itinerary of Saint-Denis's journey, taken from the map.

57. Council of Marine to Le Maire, Oct. 28, 1716, AC, B 38:326, LC transcript.

58. Saint-Denis to Cadillac, Sept. 7, 1715, extracted in Cadillac to Pontchartrain, Feb. 7, 1716, AC, C13A, 4:617–26, LC transcript.

59. Saint-Denis's first declaration.

60. Saint-Denis to Cadillac.

61. Shelby, "International Rivalry," 190–92.

62. Comte de Maurepas to Pierre de Rigaud Vaudreuil (Louisiana governor), Jan. 13, 1744, and Maurepas to Saint-Denis, same date, AC, B 78:448, LC transcript, microfilm, Center for Louisiana Studies, University of Southwest Louisiana.

Le Maire and the "Mother Map"

1. For a detailed treatment, see Jackson and Weddle, "The Oliván Rebolledo–Saint-Denis Maps of Texas, Louisiana, and New Spain, 1715–17," this volume.

2. Surveys include William Edward Dunn, *Spanish and French Rivalry in the Gulf Region of the United States, 1678–1702: The Beginnings of Texas and Pensacola;* Charmion Clair Shelby, "International Rivalry in Northeastern New Spain, 1700–1725" (Ph.D. diss., University of Texas, 1935); Lawrence Carroll Ford, *The Triangular Struggle for Spanish Pensacola, 1689–1739;* Henry Folmer, *Franco-Spanish Rivalry in North America, 1524–1763*, vol. 7 of *Spain in the West.*

3. Jay Higginbotham, *Old Mobile: Fort Louis de la Louisiane, 1702–1711*, p. 265.

4. Ibid., p. 268; Marcel Giraud, *A History of French Louisiana: The Reign of Louis XIV, 1698–1715*, trans. Joseph C. Lambert, 1:239.

5. Higginbotham, *Old Mobile*, p. 265.

6. Jean Delanglez, trans., "M. Le Maire on Louisiana," *Mid-America* 19, no. 2 (Apr., 1937): 143, 153. This is a translation of Le Maire's letter of Jan. 15, 1714, from Pensacola, with four pages of introductory material by Delanglez.

7. Ibid., pp. 153–54.

8. Higginbotham, *Old Mobile*, p. 278; Giraud, *History of French Louisiana*, 1:240.

9. Bienville to Pontchartrain, Feb. 25, 1708, in *MPA*, ed. Dunbar Rowland and Albert Godfrey Sanders, 3:120.

10. Higginbotham, *Old Mobile*, p. 278.

11. Giraud, *History of French Louisiana*, 1:241–42.

12. Higginbotham, *Old Mobile*, p. 350.

13. Giraud, *History of French Louisiana*, 1:320.

14. Delanglez, "Le Maire," p. 129. For an account of the death of two Spanish priests at Pensacola in 1711, see Ford, *Triangular Struggle*, pp. 86–87.

15. Delanglez, "Le Maire," pp. 136–37.

16. Ibid., p. 128.

17. Cadillac to Council, Jan. 2, 1716, *MPA*, 2:210–11.

18. Hubert to Council, Oct. 27, 1717, *MPA*, 2:241–42.

19. Giraud, *History of French Louisiana*, 1:320.

20. Delanglez, "Le Maire," pp. 129–30.

21. Charles Edward O'Neill, *Church and State in French Colonial Louisiana: Policy and Politics to 1732*, p. 117. Marcel Giraud, *Histoire de la Louisiane française: L'epoque de John Law, 1717–1720*, 3:132, suggests another possible reason for Le Maire's recall. Delisle was planning to update his 1718 map with Bobé's help. Le Maire's presence in Paris would have aided the revision and they may have asked Director Raudot to withdraw him for that purpose, as all three men were indebted to Le Maire for their knowledge of Louisiana. See note 35.

22. Delanglez, "Le Maire," p. 128.

23. Giraud, *Histoire de la Louisiane française*, 3:136–37.

24. Ibid., p. 360; Delanglez, "Le Maire," p. 128.

25. Giraud, *Histoire de la Louisiane française*, 3:147. After Marigny died in 1728, his widow–the niece of François Le Maire–married Ignace François Broutin, who became engineer-in-chief of Louisiana and produced many excellent maps and architectural plans for the colony. Broutin's mother was also a Le Maire, relationship to the priest unknown (Samuel Wilson, Jr., "Ignace François Broutin," in *Frenchmen and French Ways*, ed. John Francis McDermott, p. 253). Further, Marigny's son, Antoine Philippe de Marigny de Mandeville, was a skilled draftsman who drew several maps of Louisiana (Walter J. Saucier and Kathrine Wagner Seineke, "François Saucier," in ibid., p. 208). Considering that Le Maire's brother worked in the Marine Ministry, it seems that the entire clan was cartographically inclined.

26. Adrien Launay, *Memorial de la Société des Missions Étrangères, 1658–1913*, p. 386. Juthia, sometimes spelled Yuthia or Ayuthia (now Ayutthaya), was the former capital of Siam, situated on an island in the Chao Phya River, about forty-five miles north of present-day Bangkok.

27. The map is Le Maire's *Carte nouvelle . . . 1716* (old BSH, 4044C, 46) reproduced as fig. 8. This is presently in SHM, recueil 68, piece 57, reproduced from a copy at the OSMRL. The notation reads as follows: "Father Charlevoix, in his *Histoire [et description] générale de la Nouvelle France,* published Dec. 1744, Vol. I, p. 170 and following, instructs us of the voyage by land of M. de St. Denis to Mexico. See also the relation in my memoirs of this voyage of M. de St. Denys [*sic*] *translated from the Spanish in the year 1715* [emphasis added]."

28. Giraud, *Histoire de la Louisiane française,* 3:129–30.

29. Delanglez, "Le Maire," p. 137.

30. Le Maire, "Mémoire sur la Louisiane," 1718, AC, C13C, 2:153. This record group is found at the AN; copies at LC.

31. Delanglez, "Le Maire," p. 140.

32. Ibid., p. 137.

33. Ibid., p. 127.

34. Ibid., p. 125.

35. John C. Rule, "Jérôme Phélypeaux, Comte de Pontchartrain, and the Establishment of Louisiana, 1696–1715," in *Frenchmen and French Ways in the Mississippi Valley,* ed. John Francis McDermott, pp. 182–85. Marcel Giraud, in *Histoire de la Louisiane française: Années de transition, 1715–1717,* 2:15–18, 3:129–32, explores Le Maire's influence on Bobé, Raudot, and the Delisles.

36. See the Bobé-Delisle correspondence, formerly housed in Archives du Service Hydrographique de la Marine (ASH), 115x, no. 26. This record group is now at the AN, and has been assigned the number 3 JJ 387; copies at LC.

37. Ronald Vere Tooley, *French Mapping of the Americas: The De l'Isle, Buache, Dezauche Succession, 1700–1830,* lists the Delisles' various maps and gives some biographical information.

38. Giraud, *Histoire de la Louisiane française,* 2:15, 18.

39. Delanglez, "Le Maire." Delanglez, in preparing his translation for *Mid-America* 19, worked from a copy in the Ayer Collection, Newberry Library, Chicago, listed as MS. 293, 2:1–46. This letter is also found in AC, C13C, 2:109.

40. The extracts made by Bobé and Guillaume and Claude Delisle are found in Bibliothèque du Muséum d'Histoire Naturelle, Paris, MS. 948; ASH, 115x, no. 22B; and ASH, 115xxxii, no. 4, respectively.

41. MS. map 98, Ayer Collection, Newberry Library. The map is bound facing Le Maire's letter cited in note 39. It was evidently traced by the clerk who compiled this four-volume set of memoirs, based on Le Maire's original sketch of Mobile Bay (not located).

42. Le Maire, "Mémoire sur la Louisiane pour estre présenté avec la carte de ce pais au Conseil Souverain de la Marine," Mar. 7, 1717, BN, MS. fr. 12105; copy in LC. Giraud cites this document as Mar. 1.

43. Gustave Devron, ed., "François Le Maire. Extrait d'un Mémoire sur la Louisiane, 1717," *Comptes-rendus de l'Athénée Louisianais* (Oct., 1899). Important sections of the original, especially concerning Texas, are not found in this extract.

44. The Company of the West, chartered Aug., 1717, was reorganized in May, 1719, as the Company of the Indies. Folmer, *Franco-Spanish Rivalry,* p. 245.

45. Bobé's interest in the "Sea of the West" is revealed by his letters to Delisle, ASH, 115x, no. 26.

46. Le Maire, "Mémoire sur la Louisiane," 1718, AC, C13C, 2:153; copy in LC.

47. Ibid. Derbanne's journal is translated in Katherine Bridges and Winston De Ville, "Natchitoches and the Trail to the Rio Grande: Two Early Eighteenth Century Accounts by the Sieur Derbanne," *Louisiana History* 8, no. 3 (Summer, 1967): 239–52.

48. Le Maire, "Mémoire sur la Louisiane," 1718.

49. Le Maire, "Des moeurs des sauvages de la Louisiane," May 13, 1718, ASH, 67, no. 4.

50. Reference is to Le Maire's translation of González de Mendoza's *Historia de las cosas . . . de la China,* published in Madrid, 1586. See Le Maire, "Mémoire sur la Louisiane," March 7, 1717.

51. See note 27 for Le Maire's notation on his *Carte nouvelle . . . 1716.* Evidently Le Maire received a copy of Saint-Denis's 1715 declaration taken at Mexico City and forwarded a translation to France. Several versions of this declaration are in the French archives, cited in notes 67 and 100.

52. Bobé to Delisle, Jan. 15, 1710, ASH, 115xvi, no. 3.

53. The map is probably Delisle's *L'Amérique septentrionale,* 1700, where Texas is cut in half by a line dividing "New Mexico" from "Florida." Delisle's 1703 *Carte du Mexique* claims virtually all of Texas for France, so its boundary line could not have annoyed Bobé.

54. Bobé to Delisle, Jan. 15, 1710, ASH, 115xvi, no. 3. Father Francisco Hidalgo wrote to Louisiana on Jan. 17, 1711, but his letter did not reach Governor Cadillac until 1713 (Carlos E. Castañeda, *Our Catholic Heritage in Texas,* 2:27–28). The Hasinai Indians were variously called "Cenis" by the French and "Asinais" or "Tejas" by the Spaniards.

55. Giraud, *Histoire de la Louisiane française,* 3:130.

56. Le Maire, "Mémoire sur la Louisiane," Mar. 7, 1717. Giraud, 2:18–24, gives an account of the illusive Sea of the West. Claude Delisle offered his "conjectures" on the subject in 1702 (ASH, 115xi, no. 12), and Guillaume did likewise in 1717 (ibid.; also see Archives du Ministère des Affaires Étrangères, Mémoires et Documents, Amérique, 1:241). Concerning the "map and memoir" that Guillaume Delisle had presented "to prove the existence and location of the Western Sea," Bobé wrote: "I am happy that it brought you such honor, that the council savors your proposal, and that they appear to be inclined toward the exploration of that sea. I [have] the honor of telling you, Monsieur, that I have been trying for four years to convince the Court [to do so]" (Bobé to Delisle, [undated, but prior to Mar., 1717?] ASH, 115x, no. 26A). Indeed, in April of 1718, Bobé, encouraged by the interest taken by Delisle and Le Maire, penned his own "Mémoire pour la découverte de la mer de l'ouest" (AC, C11E, 16:64; also see Bibliothèque Mazarine, MS. 2006, no. 5; and Ayer MS. 293, 3:109–226, Newberry Library, Chicago). Their combined efforts resulted in Father Charlevoix being sent to Louisiana in 1720 to investigate the question, but his report did not end French obsession with the Sea of the West. It began appearing on maps at mid-century and persisted for decades. See William P. Cumming, et al., *The Exploration of North America, 1630–1776,* pp. 181–82.

57. Delanglez, "Le Maire," p. 136. The "fairy tale" refers to the spurious exploration by Louis-Armand de Lom d'Arce, Baron de Lahontan, of North America and particularly his "Rivière Longue." Lahontan's popular book and map were published in 1703.

58. Shelby, "International Rivalry," pp. 111–12.

59. See fig. 1 in Jackson and Weddle, "The Oliván Rebolledo–Saint-Denis Maps," this volume. A trip to the Río Grande by Saint-Denis as early as 1705 is doubtful because (1) he had to have Indians guide him there in 1714, (2) his reception indicates that it was his first time at the presidio, and (3) neither French nor Spanish sources contain a hint of an earlier trip.

60. Minet's two maps of La Salle's landing are reproduced in Robert S. Weddle, ed., *La Salle, the Mississippi, and the Gulf: Three Primary Documents,* plates 7 and 8; his *Carte de la Lovisiane,* thought by many authorities to have been a copy of La Salle's missing map, is in Sara Jones Tucker, *Indian Villages of the Illinois Country: Part I, Atlas,* plate VII.

61. Ronquillo's copies of Minet are in the AGI, Mapas y Planos, described in Woodbury Lowery, *The Lowery Collection: A Descriptive List of Maps of the Spanish Possessions within the Present Limits of the United States, 1502–1820,* items 186 and 187; they are reproduced in Robert S. Weddle, *Wilderness Manhunt: The Spanish Search for La Salle,* plates 3 and 6.

62. Delanglez, "The Sources of the Delisle Map of America, 1703," *Mid-America* 25, no. 4 (Oct., 1943): 298, gives locations for the various French maps that reflect Sigüenza's information. Bernou's *Carte* is reproduced in Tucker, *Indian Villages,* plate VIII. The Sigüenza map is in the AGI, Mapas y Planos, described in Lowery, item 193, and reproduced from a photostat in the BTHC. The AGI has a slightly different copy, photographed by Karpinski, and yet another variation is in the Biblioteca Nacional, Madrid.

63. Diego Peñalosa's map (from old BSH, 4049B, 28) appears in Carl I. Wheat, *Mapping the Transmississippi West, 1540–1861,* 1, opp. p. 44; the assistance that Vincenzo Coronelli received from Abbé Bernou is documented in Delanglez, "Sources of the Delisle Map of America, 1703," p. 297. Another copy of the Peñalosa map is found in Delisle's "sketch maps" at AN, 6 JJ 75, 270.

64. Cadillac to Pontchartrain, Jan. 2, 1716, AC, C13A, 4:509–35. Le Maire sent his maps directly to Bobé, and it is doubtful that the governor saw them, considering Le Maire's low opinion of him.

65. Jean-Baptiste Bénard, Sieur de La Harpe, *Historical Journal of the Establishment of the French in Louisiana,* p. 99.

66. According to a memoir of Jan. 11, 1717 (AC, C13A, 5:229–36), "many letters" had been received from Saint-Denis in Mexico. Cadillac reported only two of them to the Council: one from San Juan Bautista, the other from Mexico City. They are extracted in Cadillac to Pontchartrain, Feb. 7 and July 21, 1716, AC, C13A, 4:617–26.

67. Moro to Gallut, Sept. 20, 1715, AC, C13C, 4:50, extracted in AC, C11A, 36:415, and ASH, 115x, no. 25. In a notation on his *Carte nouvelle . . . 1716,* Le Maire says he has translated a 1715 account of Saint-Denis's journey; perhaps it was from a copy received from Moro or Saint-Denis himself.

68. Weddle, *Wilderness Manhunt,* pp. 252–65, discusses the Talons; their account of life in Texas is translated in Weddle, ed., *La Salle,* pp. 209–58, along with additional biographical information.

69. Cadillac to Pontchartrain, Oct. 26, 1713, AC, C13A, 3:1–93, extracted in *MPA,* 2:180.

70. Cadillac to Minister [Pontchartrain], July 21, 1716, AC, C13A, 4:617–26, notes receiving Saint-Denis's letter from San Juan Bautista dated Feb. 21, 1715. Shelby, "International Rivalry," p. 124, suggests that the Talons were its bearers.

71. Cadillac to Pontchartrain, Jan. 2, 1716, AC, C13A, 4:509–35. According to Folmer, *Franco-Spanish Rivalry,* p. 234, Saint-Denis did transmit a map to Cadillac.

72. This untitled map is presently in BN, Cartes et Plans, 138 *bis*-1-7. When the map was photographed by Doysié in 1921–22, it was noted as being attached to Cadillac's letter of Jan. 20 [*sic*], 1716. This copy is now in the Clements Library, University of Michigan, Ann Arbor; Karpinski photographed it as well. Émile Lauvrière, *Histoire de la Louisiane française, 1673–1939,* p. 281, in discussing Saint-Denis's activities, reproduces a facsimile.

73. On the Yatasses, see Frederick Webb Hodge, ed., *Handbook of American Indians North of Mexico,* 2:993. They had been first contacted by Saint-Denis in 1700 and were a tribe of the Caddo confederacy, linguistically related to the Natchitoches.

74. Cadillac to Pontchartrain, Jan. 2, 1716, AC, C13A, 4:509–35.

75. Cadillac, in his letter of Oct. 26, 1713, to Pontchartrain, claimed that three routes were known, the third going inland from another bay below La Salle's–probably modern San Antonio Bay or Corpus Christi Bay.

76. Le Maire, forwarding maps to the Council, Jan. 25, 1716, AN, B1 (Marine), 9:273; copy in LC.

77. Minutes of the Council, Aug. 29, 1716, AC, C13A, 4:205;copy in LC.

78. Bobé to Delisle, Aug. 17, 1716, in *Historical Magazine & Notes & Queries* 3 (Aug., 1859): 233.

79. Council to Le Maire, Oct. 28, 1716, AC, B 38:326; copy in LC. This recognition by the Council, according to Le Maire, only aroused the jealous anger of the secular authorities in Louisiana against him (O'Neill, *Church and State,* p. 115).

80. Le Maire's *Carte nouvelle,* (old BSH, 4044C, 46) is presently in SHM, recueil 68, piece 57, reproduced from a copy at the OSMRL. See also Mildred Mott Wedel, *The Deer Creek Site, Oklahoma: A Wichita Village Sometimes Called Ferdinandina, an Ethnohistorian's View,* pp. 23, 24.

81. BN, Cartes et Plans, Ge D 7883. This map was photographed by Karpinski; copies at the Newberry Library and the LC. It has been reproduced in facsimile by Gravier and Lauvrière.

82. BN, Cartes et Plans, 138 *bis*-1-6, reproduced from a copy at Newberry Library, Chicago. See Lauvrière, *Histoire,* for a facsimile reproduction.

83. The notation reads: "One should depend on this map for the interior of the lands of the Carolina coast and Bay of St. Bernard, but for the entrance to the Miciscipi as well as for Lakes Pontchartrain, Maurepas and Manchac, one should consult the two others in order to draw one that is exact."

84. AN, 6 JJ 75, 238. The full title of this map reads: *Coste de la Louisiane, partie inferieure du Miciscipi; environs du Fort Louis et de Pensacola; et partie du cours de la R. Rouge. Par F. Le Maire, P.M. ap. 1718.* There is a flap covering the lower Mississippi and the coast between Lake Manchac and Mobile Bay, perhaps for the purpose of corrections by Delisle.

85. Le Maire, "Mémoire sur la Louisiane," Mar. 7, 1717, BN, MS. fr. 12105.

86. Bobé to Delisle [undated, but prior to Mar. 1717?], ASH, 115x, no. 26A.

87. Bobé to Delisle, Mar. 8 and Aug. 2, 1717, ASH, 115x, no. 26B and 26E. The latter item mentions Le Maire's brother, a liaison of the Marine Ministry at court, who served as a conduit between Bobé at Versailles and Delisle in Paris.

88. In several of his memoirs, Le Maire speaks of Delisle maps, indicating that he used Delisle's 1700 *L'Amérique septentrionale* and/or Delisle's 1703 *Carte du Mexique*.

89. The fig. 13 map is in BN, Cartes et Plans, 138 *bis*-1-3, and is reproduced from a copy at the OSMRL. Without maker or date, it appears to be a prototype for de Fer's 1701 *Les Costes aux environs de la rivière de Misisipi* discussed in Robert Sidney Martin and James C. Martin, *Contours of Discovery: Printed Maps Delineating the Texas and Southwestern Chapters in the Cartographic History of North America, 1513–1930. A User's Guide*, pp. 38–39. See also Lowery, *Collection*, item 251.

90. Santiago de la Monclova was founded by Alonso de León in 1689, on the former site of Luis de Carvajal's Nuevo Almadén. The presidio, San Francisco de Coahuila, and the settlement served as capital of the province (Vito Alessio Robles, *Coahuila y Texas en la época colonial*, pp. 351–57).

91. Le Maire, "Mémoire sur la Louisiane," 1718, AC, C13C, 2:153.

92. *Carta de la costa de Pansacola dende [desde] Santa Rosa hasta Massacra. MDCCXIII* (unsigned), AGI, Mapas y Planos, Florida y Luisiana, 34. This map was detached from *legajo* 524, Mexico (old number 61-3-12). In his Jan. 15, 1714, letter Le Maire describes it: "I have drawn a plan of the coast and of the fort which has been sent to the King of Spain" (Delanglez, "Le Maire," p. 137).

93. Salinas also accompanied the 1690 Llanos voyage, during which Cárdenas executed his map of San Bernardo Bay. For Salinas's service record, dated June 20, 1701, see Henry R. Wagner, *The Spanish Southwest, 1542–1794: An Annotated Bibliography*, p. 307; Alessio Robles, *Coahuila y Texas*, pp. 366–69.

94. Oliván, "Informe," Dec. 24, 1717, AGI, Mexico 61-6-35 (BTHC transcript).

95. Moro to Gallut, Sept. 20, 1715, AC, C11A, 36:415.

96. Cadillac to Pontchartrain, Oct. 26, 1713, *MPA*, 2:180.

97. Salinas to the king, Jan. 20, 1717; Salinas to the viceroy, Feb. 15, 1717, AGI, Mexico 61-6-35 (BTHC transcript).

98. Salinas to the viceroy, Feb. 15, 1717, AGI, Mexico 61-6-35 (BTHC transcript). Many of Le Maire's coastal toponyms between La Salle's bay and the Mississippi suggest that the missionary also had access to Juan Enríquez Barroto's diary or a version of his map. The Sigüenza map does not extend far enough eastward to have been the source for these names.

99. Saint-Denis, declaration, Sept. 18, 1717, AGI, Mexico, 61-6-35 (BTHC transcript).

100. "Mémoire sur la Louisiane et relation du voyage du Sr. de St. Denis" [undated, but ca. Sept., 1717], AC, C13A, 4:1001. This memoir, to which is attached a copy of Saint-Denis's first declaration, contains an interesting account of the Frenchman's arrest and imprisonment in 1717. Its author, as Giraud suspects (2:187), was probably Gerardo Moro.

101. Le Maire's 1718 memoir begins: "I took the liberty last year [1717] to send this memoir and several maps to the Royal Council of Marine."

102. Bobé to Delisle, Oct. 16, 1717, ASH, 115x, no. 26F.

103. Bobé to Delisle, Mar. 18, 1718, ASH, 115x, no. 26H.

104. Although not bearing de Fer's name, this *Carte de la Nouvelle France . . . pour l'etablissement de la Compagnie Françoise Occident* is fully cited in Lowery, *Collection*, items 272, 293. Its most interesting feature is an inset

showing the Louisiana coast in detail, perhaps based on a Le Maire sketch and/or information.

105. Delanglez, "Le Maire," p. 127.

106. De Fer's *Le Cours du Missisipi ou de St. Louis* is fully cited in William P. Cumming, *The Southeast in Early Maps . . .* , p. 186. Cumming notes that de Fer's 1718 maps were "retrogressive" compared to Delisle's map of that year.

107. Delisle's 1718 map is reproduced in almost every study of North American cartography. Our fig. 14 is a re-engraving of the original, issued by Covens and Mortier at Amsterdam, ca. 1730, reproduced courtesy of the BTHC.

108. Cited in Lowery, *Collection,* p. 230.

109. Charles O. Paullin, *Atlas of the Historical Geography of the United States,* ed. John K. Wright, p. 12.

110. Wheat, *Mapping the Transmississippi West,* 1:67, 205.

111. W. Raymond Wood, "Mapping the Missouri River through the Great Plains, 1673–1895," in *Mapping the North American Plains: Essays in the History of Cartography,* ed. Frederick C. Luebke, Frances W. Kaye, and Gary E. Moulton, p. 30.

112. Martin and Martin, *Contours of Discovery,* p. 40.

113. Raymond Thomassy, *Cartographie de la Louisiane,* p. 210.

114. Raphael N. Hamilton, "The Early Cartography of the Missouri Valley," *American Historical Review* 39, no. 4 (July, 1934): 655–57.

115. Delanglez, "Le Maire," pp. 126–27.

116. Ralph E. Ehrenberg, "A Catalog of the Exhibition," in *Mapping the North American Plains,* ed. Luebke, Kaye, and Moulton, p. 184.

117. Cumming, *Southeast in Early Maps,* pp. 39, 184, 186.

118. Bobé to Delisle, Aug. 26, 1718, ASH, 115x, no. 26O, congratulating Delisle on his appointment with its salary of 1,200 *livres.*

119. Delisle to Bobé, 1718, ASH, 115x, no. 26R.

120. Ibid.

121. Le Maire to Delisle, May 19, 1719, ASH, 115x, no. 22D.

122. Cumming, *Southeast in Early Maps,* p. 184.

123. The original is in the SHM (old BSH, 4044C, 11; current citation lacking), reproduced from a copy at Newberry Library, Chicago. The title block, on the left margin of the map, has been omitted. Wheat shows the Vermale map as item 98 (1, opp. p. 63). Perhaps the best reproduction yet is in the 1893 atlas accompanying *Reproductions de cartes et de globes relatifs a la dé'couverte de l'Amérique,* plate 25.

124. Marcel Giraud to Winston De Ville, Apr. 29, 1988. Giraud, in the frontispiece of his vol. 1, reproduces another map by Sieur Vermale, crediting Du Sault.

125. Le Maire, "Mémoire sur la Louisiane," 1718, AC, C13C, 2:153.

126. If the word "return," used on the upper route, means Saint-Denis's return to Louisiana (rather than his return to San Juan Bautista), then Vermale could well have drawn his map in 1717. As noted, Saint-Denis was back in Mobile by Aug., 1716, giving Le Maire ample opportunity to show the outbound and return routes on his missing 1717 maps. If, however, "return" means the route Saint-Denis took to return to Mexico in 1716–17, Le Maire would not have known of it until Derbanne reached Mobile in November of 1717. Consequently, Vermale could not have seen or used Le Maire's map of the second route until early 1718. Whether Vermale copied a map of 1717 or 1718

is somewhat immaterial to the main point: Le Maire was the basis for the cartographic information about Texas reflected on "Vermale's" map.

127. Le Maire to Delisle, May 19, 1719, ASH, 115x, no. 22D.

128. Delanglez, "Le Maire," pp. 141–42.

129. Although Vermale remains something of a mystery, the Bobé-Delisle letters in ASH, 115x, no. 26A–26R, establish the reliance of Delisle on Le Maire's various maps sent in the years 1716, 1717, and 1718. As only two of the 1716 maps are presently available, the problem lies in trying to determine which of Le Maire's other creations were a factor. Tantalizing hints of the 1717–18 maps emerge from Bobé's correspondence: "M. Le Maire, in his last map, draws the Miciscipi only to Ouisconsing . . ." (26F, Oct. 16, 1717); "I have already asked and have asked again that M. Le Maire [the priest's brother, who worked in the Marine Ministry] lend you the maps of M. L'Abbé [the priest, François Le Maire]. I am vexed at being unable to go to Paris to have the honor of seeing you and telling you personally my thoughts on the maps of M. Le Maire . . ." (26G, Jan. 4, 1718); "That which should reach you in about a week will cure all with the reports of M. Le Maire . . ." (26H, Mar. 18, 1718); "I believe that you would do well not [to publish] your map of Louisiana [until after the ships from Louisiana arrive] because, without doubt, the letters we will receive will give us much new information that will contribute to its perfection, above all if M. Le Maire answers precisely everything I asked him" (26J, Mar. 30, 1718). Delisle, alas, did not list his sources for the 1718 map as he and his father did in 1703, or we would know more definitely which of Le Maire's items were used.

130. BN, Cartes et Plans, Ge C 5115, reproduced from a copy at the OSMRL. This map is Wheat's item 101 (1, opp. p. 70). A two-part rough sketch for the map is in BN, Cartes et Plans, Ge DD 2987-8796 and 2987-8797 (the latter of which Wheat reproduces as item 103, opp. p. 66).

131. Wheat, *Mapping the Transmississippi West,* 1:69n, 206, lists variant copies.

132. Bridges and De Ville, "Natchitoches and the Trail to the Rio Grande," p. 246. This tribe, kinsmen of the Tonkawas, probably had their village along the Brazos, not on a western branch of the Trinity as Derbanne thought.

133. Le Maire to Delisle, May 19, 1719, ASH, 115x, no. 22D.

134. *Carte nouvelle de la partie de l'ouest de la Louisianne* . . . , ca. 1725, attributed to Jean Beaurain, is discussed in Mildred Mott Wedel, "The Bénard de la Harpe Historiography on French Colonial Louisiana," *Louisiana Studies* 13, no. 1 (Spring, 1974), p. 65. The original is in LC, MS. Div., Louisiana Miscellany 215-2654, bound in Beaurain's copy of La Harpe's *Journal historique.* The map is reproduced in its entirety in *The Exploration of North America,* p. 170, and *Mapping the North American Plains,* fig. I.3. See also Wedel, "J.-B. Bénard, Sieur de la Harpe, Visitor to the Wichitas in 1719," *Great Plains Journal* 10, no. 2 (Spring, 1971), opp. p. 44. The Delisle sketches at the AN contain a preliminary draft (without legends) that bears an uncanny resemblance to the "Beaurain" map; it is 6 JJ 75, 266.

135. Wedel, "La Harpe Historiography," p. 10. Bienville to the Company of the Indies, Oct. 4, 1721, *MPA,* 3:310–11, expresses disgust with La Harpe's failure and his self-serving journal of the expedition. One important achievement, however, did result from this botched attempt: the map of Galveston Bay drawn by Valentin Devin. It is featured on his *Carte de la coste de la Louisiane* . . . , Ayer MS. map 159, and survives in the form of a plan illustrating

La Harpe's LC manuscript, another copy having made its way into the Rosenberg Library, Galveston. They are titled, respectively, *Plan du port françois ainsy nomme par M. Benard de la Harpe* . . . and *Plan du port decouvert dans le Golfe du Mexique*. . . . See also the bay labeled "Port decouvert par Mr. de la Harpe en 1720 [*sic*]" on Beaurain's *Carte nouvelle* (fig. 17), no doubt derived from Devin's chart.

Bibliography

Primary Sources

Paris

Archives Nationales (AN)

Archives des Colonies (AC)

Série B: Ordres du Roi et Dépêches de la Marine. Registers 38, 78.

Série C: Correspondance générale.

C11A–Nouvelle France. Register 36.

C11E–Canada, divers. Register 16.

C13A–Louisiane. Registers 3, 4, 5.

C13C–Louisiane. Registers 2, 4.

Archives de la Marine

Série B: Correspondance. B1, Register 9.

Série JJ: Service Hydrographique.

3 JJ–Registers 387–88 (collection of Delisle papers formerly designated 115x, and following).

6 JJ–Register 75, piece 238 (Le Maire, *Coste de la Louisiane . . . 1718* in Delisle sketch maps).

Bibliothèque Nationale (BN)

Fonds Français: 12105–Le Maire, "Mémoire sur la Louisiane," March 7, 1717.

Cartes et Plans:

Portfolio 138 *bis*-1-3–anonymous, *Carte du Mississipy,* ca. 1700.

Portfolio 138 *bis*-1-6–Le Maire, northern coast of Gulf of Mexico, 1716.

Portfolio 138 *bis*-1-7–map of Texas and Louisiana, said by Gov ernor Cadillac to have been drawn by his son, 1716.

Ge C 5115–Beauvilliers, *Carte nouvelle de la partie de l'ouest de province de la Louisiane . . . 1720.*

Ge D 7883–Le Maire, *Carte nouvelle de la Louisiane . . . 17*

Service Historique de la Marine (SHM), Vincennes

Recueil 68, piece 57 (old BSH, 4044C, 46)–Le Maire, *Cart velle de la Louisiane . . . 1716.*

(Current citation lacking; old BSH, 4044C, 11)–Vermale, *Carte générale de la Louisiane . . . 1717.*

Note: Documents from the Paris archives were consulted using copies in the Library of Congress. Maps from the Paris archives were consulted using prints held in the OSMRL, as well as photocopies made by Louis Karpinski and deposited at the Newberry Library, Chicago.

Seville

Archivo General de Indias (AGI)

Audiencia de Guadalajara 67-3-28 (old number).

Audiencia de Mexico 61-1-34, 61-2-12, 61-6-35, 62-1-41 (old numbers).

Mapas y Planos

Florida y Luisiana 34–Le Maire, *Carta de la costa de Pansacola . . . 1713.*

Mexico 5–Alvarez de Pineda, untitled chart of Gulf of Mexico, 1519.

Mexico 79, 80–Ronquillo's copies of Minet, 1687.

Mexico 86–Sigüenza, *Mapa del camino que el año 1689 hizo el Gobernador Alonso de León. . . .*

Mexico 88–anonymous, *Mapa del viaxe que el año 1690 hizo el Gobernador Alonso de León desde Coahuila hasta la Carolina.*

Mexico 89–Cárdenas, *Planta cosmográphica del lago de San Bernardo . . . 1690.*

Mexico 90–Terán, *Mapa de la provincia donde habita la nación Casdudacho . . .* , 1691.

Mexico 110–Oliván, *Mapa geográfico . . . 1717.*

Note: For the Spanish archives at Seville, transcripts at the BTHC were used. Copies of the AGI maps are available at the BTHC (Bryan Collection), the Texas State Archives (Inglis Collection), and the Newberry Library (Karpinski Collection).

Mexico City

Archivo General de la Nación (AGN)

Historia (H) 27 (BTHC transcript)

Provincias Internas (PI) 181 (BTHC transcript)

Gobernación, Mapoteca 197.2, 197.3, 197.6 (Oliván maps)

Chicago

Newberry Library, Ayer Collection

MS. 293–"Copie d'une lettre ecritte de Pensacola le 15[e] Janvier 1714, par M. Le Maire."

MS. map 98–[Le Maire,] *Fort Louis, province de la Louisiane . . .* (bound with the above document).

Karpinski Collection of maps photographed from the archives of France, Spain, and Portugal.

Delisle Sketch Maps, photocopies from the AN.

Austin
Barker Texas History Center (BTHC), University of Texas
Shelby Papers.
AGI Transcripts.
AGN Transcripts.

Washington, D.C.
Library of Congress, MS. Div., Louisiana Miscellany 215–2654 – Beaurain (attrib.), *Carte nouvelle de la partie de l'ouest de la Louisianne . . .*, ca. 1725.

Secondary Sources

Published

Alessio Robles, Vito. *Coahuila y Texas en la época colonial.* 1938. Reprint. Mexico: Editorial Porrúa, 1978.

Bolton, Herbert Eugene. *Texas in the Middle Eighteenth Century: Studies in Spanish Colonial History and Administration.* Berkeley: University of California Press, 1915.

———. "The Location of La Salle's Colony on the Gulf of Mexico," *Southwestern Historical Quarterly* 27, no. 3 (January, 1924): 171–89.

Bridges, Katherine, and Winston De Ville, eds. and trans. "Natchitoches and the Trail to the Rio Grande: Two Early Eighteenth-Century Accounts by the Sieur Derbanne," *Louisiana History* 8, no. 3 (Summer, 1967): 239–59.

Bryan, James P., and Walter K. Hanak. *Texas in Maps.* Austin: University of Texas, 1961.

Castañeda, Carlos E. *Our Catholic Heritage in Texas.* Vol. 2. Austin: Von Boeckmann–Jones Co., 1936–50.

Catálogo de Ilustraciones: Centro de información gráfica del Archivo General de la Nación. Vol. 10. Mexico City: Archivo General de la Nación, 1979–82.

Cumming, William P. *The Southeast in Early Maps with an Annotated Checklist of Printed and Manuscript Regional and Local Maps of Southeastern North America during the Colonial Period.* 1958. Reprint. Chapel Hill: University of North Carolina Press, 1962.

———, S.E. Hillier, D.B. Quinn, and G. Williams. *The Exploration of North America, 1630–1776.* New York: G.P. Putnam's Sons, 1974.

Delanglez, Jean. "M. Le Maire on Louisiana." *Mid-America* 19, no. 2 (April, 1937): 124–54.

———. "The Sources of the Delisle Map of America, 1703," *Mid-America* 25, no. 4 (October, 1943): 275–98.

Devron, Gustave, ed., "François Le Maire. Extrait d'un mémoire sur la Louisiane, 1717," *Comptes-rendus de l'Athénée Louisianais* (October, 1899).

Dunn, William Edward. *Spanish and French Rivalry in the Gulf Region of the United States, 1678–1702: The Beginnings of Texas and Pensacola*. Bulletin 1705, Studies in History I. Austin: University of Texas, 1917.

———. "The Spanish Search for La Salle's Colony on the Bay of Espíritu Santo, 1685–1689," *Southwestern Historical Quarterly* 19, no. 4 (April, 1916): 323–69.

Ehrenberg, Ralph E. "Exploratory Mapping of the Great Plains before 1800" and "A Catalog of the Exhibition," in Frederick C. Luebke, Frances W. Kaye, and Gary E. Moulton, eds., *Mapping the North American Plains: Essays in the History of Cartography*. Norman: University of Oklahoma Press with the Center for Great Plains Studies, University of Nebraska, Lincoln, 1987.

Folmer, Henry. *Franco-Spanish Rivalry in North America, 1524–1763*. Vol. 7 of *Spain in the West*. Glendale, California: Arthur H. Clark Co., 1953.

Ford, Lawrence Carroll. *The Triangular Struggle for Spanish Pensacola, 1689–1739*. Washington, D.C.: The Catholic University of America Press, 1939.

Giraud, Marcel. *Histoire de la Louisiane Française: Années de transition, 1715–1717*. Paris: Presses Universitaires de France, 1958.

———. *Histoire de la Louisiane française: L'époque de John Law, 1717–1720*. Paris: Presses Universitaires de France, 1966.

———. *A History of French Louisiana: The Reign of Louis XIV, 1698–1715*. Translated by Joseph C. Lambert. Baton Rouge: Louisiana State University Press, 1974.

Gravier, Henri. *La colonisation de la Louisiane a l'époque de Law. Oct. 1717–Janv. 1721*. Paris, 1904.

Hamilton, Raphael N. "The Early Cartography of the Missouri Valley," *American Historical Review* 39, no. 4 (July, 1934).

Higginbotham, Jay. *Old Mobile, Fort Louis de la Louisiane, 1702–1711*. Mobile, Alabama: Museum of the City of Mobile, 1977.

Historical Magazine & Notes & Queries 3 (August, 1859).

Hodge, Frederick Webb, ed., *Handbook of American Indians North of Mexico*. Vol. 2. Bureau of American Ethnology Bulletin 30. Washington, D.C.: Smithsonian Institution, 1907, 1910.

La Harpe, Jean-Baptiste Bénard, Sieur de. *Historical Journal of the Establishment of the French in Louisiana*. Edited by Glenn Conrad; translated by Virginia Koenig and Joan Cain. USL History Series, no. 3. Lafayette: University of Southwestern Louisiana, 1971.

Launay, Adrien. *Memorial de la Société des Missions Étrangères, 1658–1913*. Paris: Société des Missions Étrangères, 1916.

Lauvrière, Émile. *Histoire de la Louisiane française, 1673–1939*. Romance Languages Series, no. 3. Baton Rouge: Louisiana State University Press, 1940.

Lowery, Woodbury. *The Lowery Collection: A Descriptive List of Maps of the Spanish Possessions within the Present Limits of the United States, 1502–1820*. Edited by Philip Lee Phillips. Washington, D.C.: U.S. Government Printing Office, 1912.

Marcel, Gabriel. *Reproductions de cartes et de globes relatifs a la découverte de l'Amérique. Atlas.* Paris: Ernest Leroux, 1893.

Martin, Robert Sidney, and James C. Martin. *Contours of Discovery: Printed Maps Delineating the Texas and Southwestern Chapters in the Cartographic History of North America, 1513–1930. A User's Guide.* Austin: Texas State Historical Association with the Center for Studies in Texas History, University of Texas at Austin, 1982.

Navarro García, Luis. *Don José de Gálvez y la comandancia general de las provincias internas del norte de Nueva España.* Seville: Escuela de Estudios Hispano-Americanos de Sevilla, 1964.

O'Neill, Charles Edward. *Church and State in French Colonial Louisiana: Policy and Politics to 1732.* New Haven, Connecticut: Yale University Press, 1966.

Paullin, Charles O. *Atlas of the Historical Geography of the United States.* Edited by John K. Wright. Washington, D.C.: Carnegie Institution of Washington and the American Geographical Society of New York, 1932.

Pichardo, José Antonio. *Pichardo's Treatise on the Limits of Louisiana and Texas.* Vol. 1. Translated by Charles Wilson Hackett, Charmion Clair Shelby, and Mary Ruth Splawn; edited and annotated by Charles Wilson Hackett. Austin: University of Texas Press, 1931.

Priestley, Herbert I. *José de Gálvez, Visitor-General of New Spain, 1765–1771.* Berkeley: University of California Press, 1916.

Rowland, Dunbar, and Albert Godfrey Sanders, eds. and trans., *Mississippi Provincial Archives.* Vol. 2, *French Dominion, 1701–1729* (1929). Vol. 3, *French Dominion, 1701–1743* (1932). Jackson: Press of the Mississippi Department of Archives and History.

Rule, John C. "Jérôme Phélypeaux, Comte de Pontchartrain, and the Establishment of Louisiana, 1696–1715," in John Francis McDermott, ed., *Frenchmen and French Ways in the Mississippi Valley.* Urbana: University of Illinois Press, 1969.

Saucier, Walter J., and Kathrine Wagner Seineke, "François Saucier, Engineer of Fort de Chartres, Illinois," in John Francis McDermott, ed., *Frenchmen and French Ways in the Mississippi Valley.* Urbana: University of Illinois Press, 1969.

Shelby, Charmion Clair. "St. Denis's Declaration Concerning Texas in 1717," *Southwestern Historical Quarterly* 26, no. 3 (January, 1923): 165–83.

———. "St. Denis's Second Expedition to the Río Grande, 1716–1719," *Southwestern Historical Quarterly* 27, no. 3 (January, 1924): 190–216.

Thomas, Alfred Barnaby, ed. and trans., *After Coronado: Spanish Exploration Northeast of New Mexico, 1696–1727.* Norman: University of Oklahoma Press, 1935.

Thomassy, Raymond. *Cartographie de la Louisiane.* New Orleans: R. Thomassy, 1859. Separately published extract from *Géologie pratique de la Louisiane* (New Orleans, 1859), pp. 205–26.

Tooley, Ronald Vere. *French Mapping of the Americas: The De l'Isle,*

Buache, Dezauche Succession, 1700–1830. Map Collectors' Series, no. 33. London: Map Collectors' Circle, 1967.

Tucker, Sarah Jones. *Indian Villages of the Illinois Country: Part I, Atlas*. Illinois State Museum Scientific Papers. Vol. 2, no. 1. Springfield: State of Illinois, 1942.

Wagner, Henry R. *The Spanish Southwest, 1542–1794: An Annotated Bibliography*. Albuquerque: Quivira Society, 1937.

Weddle, Robert S. *Wilderness Manhunt: The Spanish Search for La Salle*. Austin: University of Texas Press, 1973.

———, ed. *La Salle, the Mississippi, and the Gulf: Three Primary Documents*. College Station: Texas A&M University Press, 1987.

Wedel, Mildred Mott. "J.-B. Bénard, Sieur de la Harpe: Visitor to the Wichitas in 1719," *Great Plains Journal* 10, no. 2 (Spring, 1971): 37–70.

———. "The Bénard de la Harpe Historiography on French Colonial Louisiana," *Louisiana Studies* 13, no. 1 (Spring, 1974). Special issue.

———. *The Deer Creek Site, Oklahoma: A Wichita Village Sometimes Called Ferdinandina, an Ethnohistorian's View*. Series in Anthropology no. 5. Oklahoma City: Oklahoma Historical Society, 1981.

Wheat, Carl I. *Mapping the Transmississippi West, 1540–1861*. Vol. 1. San Francisco: Institute of Historical Geography, 1957.

Wilson, Samuel, Jr. "Ignace François Broutin," in John Francis McDermott, ed., *Frenchmen and French Ways in the Mississippi Valley*. Urbana: University of Illinois Press, 1969.

Wood, W. Raymond. "Mapping the Missouri River through the Great Plains, 1673–1895," in Frederick C. Luebke, Frances W. Kaye, and Gary E. Moulton, eds., *Mapping the North American Plains: Essays in the History of Cartography*. Norman: University of Oklahoma Press, with the Center for Great Plains Studies, University of Nebraska, Lincoln, 1987.

Unpublished

Murphy, Henrietta. "Spanish Presidial Administration as Exemplified by the Inspection of Pedro de Rivera, 1724–1728." Ph. D. diss., University of Texas at Austin, 1938.

Shelby, Charmion Clair. "St. Denis's Second Expedition from Louisiana to the Río Grande, 1716–1719." Master's thesis, University of Texas at Austin, 1927.

———. "International Rivalry in Northeastern New Spain, 1700–1725." Ph.D. diss., University of Texas at Austin, 1935.

Index

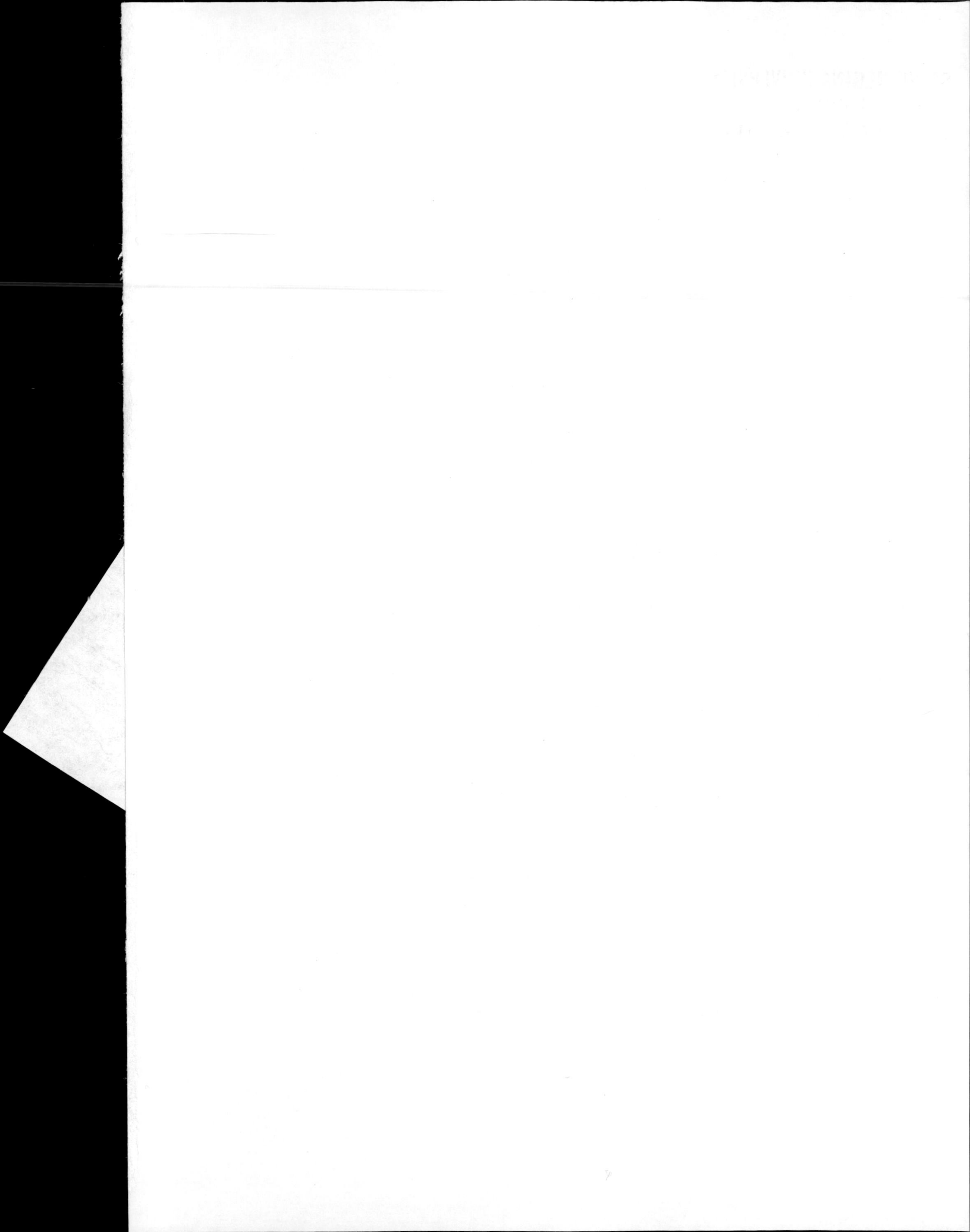

SALVE REGINA UNIVERSITY
3 3759 00155270 4

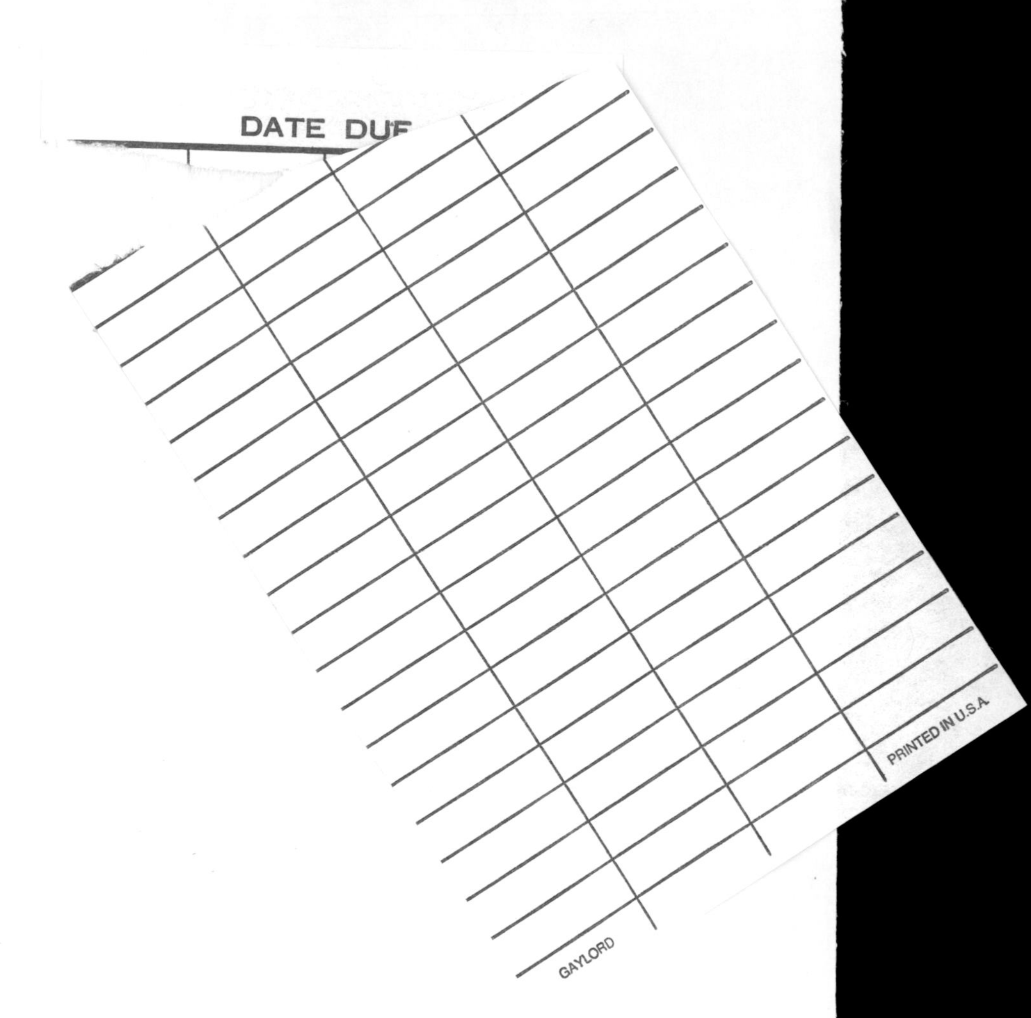
DATE DUE
PRINTED IN U.S.A.
GAYLORD